D0252815

THE
DEPTHS
OF SPACE

THE DEPTHS OF SPACE

THE STORY OF THE PIONEER PLANETARY PROBES

Mark Wolverton

Joseph Henry Press
Washington, D.C.

Joseph Henry Press • **500 Fifth Street, NW** • Washington, DC 20001

The Joseph Henry Press, an imprint of the National Academies Press, was created with the goal of making books on science, technology, and health more widely available to professionals and the public. Joseph Henry was one of the founders of the National Academy of Sciences and a leader in early American science.

Any opinions, findings, conclusions, or recommendations expressed in this volume are those of the author and do not necessarily reflect the views of the National Academy of Sciences or its affiliated institutions.

Library of Congress Cataloging-in-Publication Data

Wolverton, Mark.
    The depths of space : the Pioneer planetary probes / Mark Wolverton.
        p. cm.
    Includes bibliographical references and index.
    ISBN 0-309-09050-4 (hardcover)
    1. Pioneer Project. 2. Jupiter probes. 3. Jupiter (Planet)—
Exploration. 4. Space flight to Jupiter. I. Title.
    QB661.W75 2004
    629.43'54—dc22
                                    2004003501

All photos are reprinted courtesy of NASA, except for the photo on page 186, which is reprinted courtesy of NASA/JPL-Caltech.

Printed in the United States of America.

To my mother and father, who made my own journeys possible

Contents

Foreword   ix
    By James A. Van Allen

Acknowledgments   xi

Introduction—Message in a Bottle   1

1   Embarkation   7

2   Reaching into the Void   19

3   Something Man Has Never Done Before   40

4   The Sole Selection   52

5   Countdown and Controversy   71

6   Spring at the Cape   84

7   Twelve Generations from Galileo   100

8   Filling in the Gaps   120

9   A Jewel in the Night   138

10 Planet of Clouds   159

11 Whispers Across the Abyss   179

12 Lone Survivor   202

Notes   227

Bibliography   237

Index   241

Foreword

This book is a tale of human achievement—indomitable, perhaps heroic—at the boundary between the possible and the impossible. It carries the reader along with the gusto and fascination of a good novel. But it is not fiction. On the contrary, it is based on the author's exhaustive winnowing of primary documents and recorded interviews with numerous participants in the legendary sequence of NASA's Pioneer space missions that began in the early 1960s. It is an intensely human story centered on the extraordinary professional qualities of the project's gifted and dedicated space scientists and engineers, particularly its long-term leader, project manager Charles F. Hall. The author gives special emphasis to the far ranging missions of *Pioneer 10* and *Pioneer 11* to the planets Jupiter and Saturn and the outer reaches of the solar system. Even the participants (like me) will find here a fresh perspective on the process of scientific discovery at the frontiers of human knowledge. As a reader, one has the hopeful expectation that these little spacecraft will survive impending perils, but can scarcely wait to find out how.

The science that came out of the Pioneer's missions was both voluminous and groundbreaking, and resulted in over 500 original research papers. All of them share the dry (some say stodgy) impersonal style considered proper for professional exposition. But Mark Wolverton, a seasoned journalist, does not labor under such a constraint. He breathes life into the daily interplay of strong-minded individuals and governmental organizations, the politics and the judgment calls, the drama and the excitement of doing something for the first time, the rigors of meeting the inexorable schedule of planetary motion, the bugaboos of cost overruns and unexpected technical failures of inanimate equipment. It's all there in Wolverton's lively and perceptive words.

This is a book for anyone who seeks the vicarious experience of what it takes to pull off a space mission that has no precedent in human experience. Even though I knew much of the content from personal experience, I read every word in two 3-hour sessions, with a break for lunch, and I gained a fresh appreciation of why I have devoted my professional life to the exploration of outer space.

It's a good read, dramatic and authentic, with an engaging sense of movement and a merciful lack of tedious detail.

James A. Van Allen
University of Iowa

Acknowledgments

When I began studying the Pioneer project while on a science history fellowship at NASA Ames Research Center in the summer of 1999, I had no greater ambitions than doing an article or two. Certainly, I harbored no intention of writing an entire book on the subject. But as I met and talked at length with many of the people who had worked on Pioneer, two things became clear: first, there was much more of a story here than I'd imagined; and second, the achievements of Pioneer and its people had been all but forgotten, not just by the public but to a great degree within NASA itself. And the Pioneer veterans were keenly aware of the way in which the project had become a mere footnote in the history of space exploration. More than one interview I conducted ended with my interviewee expressing a wistful hope that "somebody would do a book about all this someday."

"Someday" came sooner than I expected, and I found myself using the several articles on Pioneer that I'd published in various magazines as a starting point for this book. Not only had I uncovered too much information that seemed to be largely ignored, I heard too many stories that I thought should be told, beginning with the experiences and reminiscences of those who worked on the project. Those constitute the heart of this book. Without exception, I found the Pioneer veterans to be generous with their time and their memories. My sincere thanks goes to all whom I had the privilege to meet and interview: Eric Burgess, Ricardo Campo, Jack Dyer, Alfred Eggers, Richard Fimmel, Tom Gehrels, Bob Jackson, Larry Lasher, Dave Lozier, Frank B. McDonald, Bruce McKibben, Bernard J. O'Brien, Robert Ryan, Robert Soberman, Anthony Tuzzolino, and Fred Wirth.

I also relied heavily on primary source material, much of it in the John Simpson papers in the Special Collections Department of the University of Chicago's Regenstein Library, where Jay Satterfield and the rest of the staff cheerfully indulged my endless requests for more boxes to search through and more copies to take home. Librarian Dan Pappas provided similar assistance in my perusal of the Ames History Collection in the center's library archives. Lynn Albaugh and John Bluck of Ames helped obtain illustrations. TRW press officer Brooks McKinney provided useful materials and led me to project personnel from the TRW side of the fence. Don Hunten, Richard Fimmel, Larry Lasher, Dave Lozier, Norman Ness, and Edward J. Smith reviewed the manuscript and offered invaluable comments and corrections. My thanks to all.

I am especially indebted to Pioneer principal investigators James Van Allen and John Simpson. As the founding father of American space science, Dr. Van Allen provided his unique perspective on Pioneer's history, and contributed the foreword.

The late Dr. Simpson was my initial guide to the history of Pioneer and was the first person to suggest to me the notion of doing a book on the subject. In a real sense, he can be considered the godfather of this book.

Thanks are also due to my agent, Michael Psaltis, for believing in and championing this project from the beginning; my perspicacious and patient editor, Jeffrey Robbins; production editor Dick Morris; and all at Joseph Henry Press.

Anne Burri Osterman, Glenn E. Bugos, Jeff Harris, Kristina Finan, Denise Shubin, Leila Monaghan, Judy Weightman, Nancy Shepherdson, Mechthild Hart, Christie Henry, Jodi Weisberg, and Cheryl Wilmeth all provided this author with indispensable support in both tangible and intangible ways. My family is an inexhaustible source of encouragement and enthusiasm both for my work and in my life.

Ames historian Glenn Bugos and I conducted what unfortunately turned out to be Charlie Hall's last interview, only weeks before his passing in 1999. I consider myself extremely fortunate to have had the opportunity to meet this extraordinary man, and can only hope that I've done him and the rest of the Pioneer team justice in these pages.

To unpathed waters, undreamed shores.

—William Shakespeare, *The Winter's Tale*, Act iv, Scene 4

Cannot you see how that little argosy will go glittering up into the sky, twinkling and glittering smaller and smaller, until the blue swallows it up. . . . It is as if a great window opened.

—H.G. Wells, *The World Set Free*

Introduction—
Message in a Bottle

No one knows the name of the first person who had the idea of tossing a message to the mercies of the ocean in an attempt to be rescued from an island or even just to make contact with a stranger in some unknown, distant land. Perhaps it was Alexander Selkirk, the 18th-century Scottish castaway who became the inspiration for Robinson Crusoe. More likely it happened much earlier, when a Greek, Phoenician, or Egyptian mariner scrawled a message on a piece of papyrus, stuffed it into a flask he'd earlier drained of wine, and flung it from the bow of his ship, wondering if it would ever reach the mysterious lands rumored to exist far across the sea.

One thing is certain, though. Until March 1972 no one had ever tried to reach across the sea of space with a physical message. Ancient peoples had carved huge pictographs and built enormous temples to communicate with their gods; people prayed, wailed, and pleaded to the skies for salvation and mercy; early in the 20th century, some who believed Percival Lowell's newspaper stories about a dying civilization on Mars proposed burning down sections of forest to form words

1

and mathematical symbols to contact the Martians. But no one, even during the first 15 years of the space age, had actually put a tangible, physical message on a spacecraft.

At first blush the idea seemed perfectly ludicrous. By the time we actually began exploring space, we knew that Lowell's Martians didn't exist, and we knew enough about the other worlds of our solar system to realize that there was little if any chance that intelligent life existed on them. Even among the small community of scientists who seriously argued for the existence of extraterrestrial civilizations, any thoughts of communication were limited to radio signals because nothing else could cross the void. Certainly no human astronauts would be leaving the solar system anytime soon, and neither would any spacecraft.

At least not until *Pioneer 10*. Aside from being the first spacecraft to pass beyond Mars, cross the asteroid belt, and travel to Jupiter, *Pioneer 10* would be moving fast enough that its velocity, coupled with a boost from Jupiter's gravity and orbital momentum, would hurtle it out of the solar system into interstellar space. It would be the fastest object ever built by humans.

Even at almost 27,000 miles an hour (7.5 miles a second), it would still take *Pioneer 10* tens of thousands of years to reach our nearest stellar neighbor, Proxima Centauri, at 4.3 light years away—assuming the craft was aimed in that direction, which it was not. But whatever trajectory *Pioneer 10* followed and however many thousands or millions of years might go by, it would someday pass near other stars, possibly other solar systems—and just maybe other solar systems populated with beings capable of intercepting a spacecraft from another star. So, since *Pioneer 10* was going anyway, why not send it off with some kind of calling card for those hypothetical aliens?

Obviously it would be a one-way message. In the unlikely event that an alien civilization ever found *Pioneer 10*, it would happen so far into the future that humans might have long since passed into extinction and so far away that any reply would be out of the question. Or humans and their technology would have long outstripped the humble origins of *Pioneer 10*, and we would have already visited the extraterrestrials ourselves. Maybe, instead of alien spacefarers, it would be the humans of the distant future who would someday find the drifting *Pioneer 10*, encountering it as an ancient artifact of a dead civilization.

So a message carried by *Pioneer 10* wouldn't be a communication in the strict sense of the word. It would be a declaration, humanity scrawling its name across the face of the universe to proclaim that we were here, we existed at least for a while, and look at what we could do. It would be a time capsule, buried in deep time and without any specific opening date prescribed. It would be a message to the future and to the present because its existence expressed faith that there would *be* a future.

When *Pioneer 10* left Earth forever on the evening of March 2, 1972, it carried one item that wasn't part of the complement of 11 scientific instruments, didn't involve the operation of the spacecraft, and didn't concern its communication with Earth like the 9-foot-diameter parabolic antenna. Mounted on the struts holding the high-gain communications antenna to the body of the spacecraft was a thin 6 × 9 inch aluminum plaque, plated with gold. The plaque was engraved with a stylized astronomical map, binary symbols, even a simple diagram of the Pioneer spacecraft.

But perhaps the most striking image on the plaque, at least to human if not ostensibly any alien eyes, was the depiction of two figures: a male and a female human being. And in an effort to be scientifically accurate, each of the human creatures was depicted in the natural state of nudity. For some people on Earth that was of much more significance than the fact that for the first time humanity was sending a calling card to the universe. Even for those who were unconcerned by this eventual interstellar display of nudity, however, the Pioneer plaque quickly became a symbol with an importance far beyond anything its creators had expected—much like the spacecraft itself.

Over 30 years after their launch, many people still remember *Pio-*

neer 10 and its sister craft, *Pioneer 11*, not as the first spacecraft to visit Jupiter and Saturn but as the probes with the plaques—the plaques with the naked people on them. It's hardly a fair or dignified manner in which to be memorialized, but perhaps oddly appropriate. In many ways the plaque carried by Pioneer into the galaxy is an exemplar of the project that produced it: simple in concept, humble in execution, yet with an impact and influence far beyond its original goals.

The Pioneer space probes are the trailblazers of the space age. They were the first spacecraft to probe the secrets of the Sun, the asteroid belt, the giant planets Jupiter and Saturn, and the void beyond Pluto. They paved the way to deep space and made possible the later successes of Voyager and Galileo. The Pioneer project drove the perfection of the communications and telemetry systems and techniques for deep-space missions and inspired the development of new technologies used not only for other deep space probes but for Earth-orbiting satellites. *Pioneers 10* and *11*, the first probes to Jupiter and Saturn, proved the concept of the gravity assist trajectory technique, indispensable for outer solar system missions.

From the beginning the philosophy of Pioneer project manager Charles F. Hall at the National Aeronautics and Space Administration (NASA) Ames Research Center in California and spacecraft builders Space Technology Laboratories (later TRW) emphasized the goals of keeping things simple and focused. New and untested design concepts would be avoided, not only because they might be unreliable but also because they were invariably expensive. Why load your spacecraft down with a lot of unnecessary frills that might not work and thus threaten the mission when you could build a craft to do a small number of tasks exceedingly well? It was an inspired notion and a spectacular success: the solar-orbiting *Pioneer 6*, launched in 1965 and designed to last only 6 months, earned the title of humanity's longest functioning spacecraft when it was contacted 35 years later in 2000.

When Daniel Goldin took the helm as NASA administrator in

1992, he introduced a new phrase to the culture of high-stakes space exploration: "faster, better, cheaper." No more would we see the spectacular but multibillion-dollar extravaganzas of a Voyager, the endless delays and escalating budgets of a Galileo, or, worst of all, the humiliating loss of billions of taxpayer dollars with the disappearance of a *Mars Observer* mission into the trackless void.

At the time much was made of Goldin's "new approach." But it was hardly new. Charlie Hall and his Pioneer team had been operating that way long before Dan Goldin came up with a snappy phrase for it. And despite the "faster, better, cheaper" philosophy's grand successes such as the *Mars Pathfinder*, as well as its notable failures such as the *Mars Polar Lander*, Pioneer did it all better. Even considering and comparing the political and budgetary realities of Pioneer's time with those of the Goldin era, the Pioneer project, dollar for dollar, spacecraft for spacecraft, and mission for mission, is still the most efficient, cost-effective, longest-lived, and most successful series of deep-space probes ever created.

The history of Pioneer is more than simply a case study in ingenious engineering and brilliant management, however. It is also a very human story of the exceptional people who made it possible — the scientists, engineers, administrators, and technicians who conceived, built, and operated the spacecraft, studied and interpreted the data, and fought the budgetary and political battles crucial to keep the project running. It is a scientific saga of pure discovery, of sleepless nights spent poring over reams of data no one had ever seen before, of frantic effort, desperate gambles, inspired ingenuity, and plain hard work.

Finally, it is a tale of surprises, of continually realizing the unexpected. Days before the 1972 launch of *Pioneer 10*, a United Press International report noted that "by the time the probe approaches the orbit of Uranus in 1979, [it] will be out of communications range of earth and will continue as a silent derelict of space." Certainly that was the expectation of nearly all who had built and worked on Pioneer, and it would mean that the spacecraft had more than accomplished its mission.

It didn't turn out that way. Over 20 years later, in the summer of 1999, *Pioneer 10* was far beyond Uranus—far beyond Pluto, in fact. Yet it was not quite a "silent derelict of space." The craft was low on power, with most of its instruments turned off, but as project manager Larry Lasher confirmed, "The signals are weak, but they're still coming."

Arthur C. Clarke once said that "the only way to define the limits of the possible is to push beyond them into the impossible." *Pioneer 10* and its brethren prove Clarke's words. Or as the *New York Times* observed after *Pioneer 10* reached Jupiter, "Despite this planet's troubles, there is very little beyond the reach of man."

This is how it happened.

1

Embarkation

I n 1959 nobody would ever have called Charles Frederick "Charlie" Hall a visionary, least of all the man himself. If anyone had ventured to suggest it, Hall would probably have laughed in his face and told the guy he was full of some unpleasant and smelly substance. Fascinated by airplanes and aviation since age 12 when he won his first airplane ride in a model airplane contest, Hall was a talented, resourceful, and creative engineer, without question. In his 18 years with NACA, the National Advisory Committee on Aeronautics, Hall contributed vital research, improving the designs of the P-38 and P-51 fighter planes, working on wing shape and design, and helping to develop the idea of conical camber—basically, a gentle curve in the surface of a wing from root to tip, an innovation that vastly improved the performance of jet aircraft.

Since graduating cum laude from the University of California at Berkeley in 1942 with an aeronautical engineering degree, Hall had worked at Ames Aeronautical Laboratory, a complex of drab government-issue buildings, hangars, and wind tunnels nestled against the southern tip of San Francisco Bay, next to the Navy's Moffett Field air base. Although to the casual visitor Moffett Field

Charlie Hall prepares a model for testing in an Ames wind tunnel, circa 1959.

was impressive enough with its enormous airship hangars and long runways, Ames wasn't much to look at. The big wind tunnels, with their twisting tubular configurations like some kind of modernistic technological sculpture, might excite some interest, but since all they did was sit there and make noise, and nobody but the engineers really understood what they did or how they worked, after the first curious glance they quickly became pretty boring.

In fact, Ames in general was pretty unglamorous. Yes, aside from the whining wind tunnels, it hosted a certain amount of flight test work, but nothing really exciting like the multi-Mach jet planes and rocketships that routinely punctured the sky in the high desert 500 miles south at Edwards Air Force Base in Southern California. At Ames they still tested aircraft with *propellers*, of all things. The laboratory also did work on helicopters, rotorcraft, and vertical takeoff and landing aircraft, all of which was marginally more interesting but hardly as sexy as the X-15 winged spaceship flying at Edwards.

To be fair, Ames had made a lot of important and respected contributions to aeronautics and its infant sister, aerospace, perhaps the most far-reaching of which was the blunt-body concept for reentry vehicles. A huge problem in space travel was how to design a vehicle to survive the searing atmospheric friction of reentry. Ames engineer Harvey Allen realized that the traditional, commonsense notion of pointing the tip of rockets and missiles to reduce aerodynamic drag was precisely the wrong idea. Allen and colleague Alfred Eggers saw that it would be much better to curve the tip of a vehicle reentering the atmosphere, thus dispersing the heat of reentry over a much wider area, while coating the vehicle surface with an ablative material to carry off heat. The idea was originally developed for the design of nuclear warheads, but it also became the critical breakthrough that allowed every astronaut from Alan Shepard to Space Shuttle crews to get home safely. For a born engineer like Charlie Hall, this kind of innovative atmosphere made Ames the only place to be.

When NACA was absorbed into NASA, the National Aeronautics and Space Administration, the staid Ames Aeronautical Laboratory (which had opened its doors in 1939 to become NACA's second lab, after the Langley Aeronautical Laboratory in Virginia) became the Ames Research Center. The name change didn't make much difference to Ames residents at first. NACA's mandate since its founding in 1915 was to keep the very lowest of profiles. Its scientists and engineers did solid research, meticulously reviewed and vetted, published in the most respected journals of the profession, all while staying decidedly out of the limelight. That also meant that NACA was completely nonpartisan: it didn't get involved in politics, in formulating policies, or even in actually building aircraft. It was a pure research organization, aloof from any considerations of influence or possible accusations of political favouritism. If you were an aircraft manufacturer who'd run into a tricky problem designing your latest fighter plane for the Navy, NACA engineers could solve your problem with their wind tunnels and slide rules, but it would be up to you to actually build the finished product and sell it to the Navy.

NASA was a different ball game from the beginning. Yes, NASA would do all kinds of research, both in aeronautics and space as its name implied, but the agency also had a *mission.* Its goal was to put America first in space, no matter what it took. Whereas NACA had striven for over four decades to stay well outside the battleground of politics, NASA was a wholly political animal from its very birth, which meant that the old NACA hands who now suddenly found themselves at NASA had to learn an entirely new way of operating, in every area from getting budgets and projects approved, to hiring and firing, to even something as mundane as ordering stationery.

With America now embroiled in the do-or-die space race, Ames found itself in something of a quandary. Its older brother, Langley, had no problem quickly assuming a leading role in the space effort, helped largely by its convenient proximity to Washington, D.C. Langley became the home of Robert Seamans's Space Task Group, the stellar assemblage of scientists and engineers charged with the job of finding the fastest way to put an American into space. Other new NASA centers, such as the Marshall Space Flight Center in Alabama and the Lewis Research Center in Ohio, also quickly found themselves in the thick of things. Meanwhile, Goddard in Maryland and the Jet Propulsion Laboratory in Pasadena were the nerve centers of America's unmanned space program, building Earth satellites and sending probes to the Moon and beyond. Ames, in contrast, was in the boondocks of Northern California, far from the action and the centers of political influence, and was known mainly for its collection of wind tunnels.

No one doubted that people such as Charlie Hall and his gifted colleagues had important contributions to make to the nation's space effort. Ask anyone at NASA headquarters or the other centers, though, and most likely they'd say that Ames might come up with some interesting ideas in materials research or vehicle shapes or even a heat shield but probably nothing more exciting than that. Even Smith DeFrance, the eminent engineer and former test pilot who had run Ames since its inception in 1939, was skeptical. A nationally respected researcher and administrator who epitomized the purely theoretical stance of NACA, DeFrance was restrained, practical, and not at all given to flights of fancy involving outlandish ideas such as space flight. Yes, Ames would do its part for NASA but not at the expense of turning the place into a circus and destroying its reputation as an august establishment for aeronautical research. This space business might be nothing more than a fluke, a curiosity, without any long-term future. There would always be airplanes and always a need to build better ones.

Such attitudes among both its own personnel and others at NASA threatened to leave Ames in the dust, as reflected by the amount of NASA's budget that went to Ames—very little, compared to the sums flowing into places like Cape Canaveral, Langley, and Huntsville, where space capsules and rockets were being created. At all the other NASA centers, new people were being hired, new facilities were being built, things were *happening*. At Ames about the only noticeable difference was that the quaint winged NACA logos had been removed from most of the buildings and wind tunnels and replaced by the striking blue NASA "meatball" logo. Things just happened more slowly there. It wasn't until a year after the changeover from NACA to NASA that DeFrance undertook a major reorganization of Ames, creating new divisions and reassigning people to bring the place more in line with the new NASA space culture.

Whatever Ames did, it would probably be little more than a footnote in America's quest for the Moon. Ames had always been known as a quiet research outpost, a place where ideas and theories were explored, not where new and unheard-of machines such as spacecraft were built. Certainly no one was thinking about Ames as the center for a major space project, whether manned or unmanned. To run such an enterprise, you needed a visionary, someone who could see the big picture, not just some engineer whose forte was details and problem solving, someone like Charlie Hall.

But there were still some people who had other ideas about space and what Ames could do in this brand new endeavor. Thanks to them,

Charlie Hall's life was about to change forever, and he didn't even know it yet.

SOMETHING USEFUL

Alfred Eggers was brazen, energetic, and, some thought, downright abrasive. Whatever you thought of him, there was no doubt that he was an idea man, one of those guys who had so many crazy notions that he didn't know what to do with them all. One, of course, had been the blunt-body concept, which he'd developed with Harvey Allen; he'd also done a lot of work on lifting bodies, the bulbous wing-less craft that ultimately led to the Space Shuttle. With the Ames reorganization in late 1959 though, Eggers found himself in charge of something called the Vehicle Environment Division, one of the new branches of Ames dedicated to NASA's space ambitions. Ames had finally begun casting about for ways to get involved in space, but as usual Eggers had bigger ideas. He remembered: "NASA Ames had never really taken on major project leadership responsibilities anywhere in the space program. We were doing lots of work useful to contributing to space activities, vehicles, trajectories, but we had nothing in the way of management responsibility." Ames had never been a place where projects were run—instead, it solved problems for projects being conducted elsewhere. Eggers wanted to change that, and he saw the arena of space as a prime opportunity. "I thought, if we had something useful to do taking leadership, we ought to have an open mind regarding getting into that role. But the first thing you gotta do is figure out what might be useful."

Space was a wide-open frontier in 1959-1960, and everything was brand new, mostly because so little was really known. If the possibilities were endless, including grandiose schemes for atomic-powered spaceships to Mars and Jupiter, huge majestic space stations, and cities inside lunar craters, so were the unknowns. No one really knew just how humans could survive in outer space. No one knew how to design spacecraft to venture beyond the Earth's orbit: What about radiation? Would such craft be pummeled by micrometeorites? How could spacecraft be controlled at such distances? How could you communicate with a craft over millions of miles of space? Aside from such practical questions, many more fundamental questions about the space environment remained. The Van Allen radiation belts had been mapped, at least partially, but what about the other planets? Did they also have radiation belts, and how would they affect visiting spacecraft? How would the Sun affect spacecraft beyond Earth's influence?

For Eggers such questions were a starting point. At the time his attitude was, "We ought to try and go places where new physics or new phenomena can be discovered which may be helpful to us in terms of understanding the laws of physics and doing things on Earth and in space." One possibility was obvious. "The one place there are certainly extraordinary physical phenomena occurring that's fairly nearby, relatively speaking, is the Sun. Getting out there, close to the Sun with proper instrumentation to acquire data, seemed to me something that should be a primary target."

Eggers called together about a dozen of his top engineers in the Vehicle Environment Division (VED), including his new deputy, Charlie Hall. Fresh from several years in charge of the 6 × 6 foot supersonic wind tunnel at Ames, Hall had been named Eggers's assistant chief upon the establishment of the VED. It was a new relationship for two old friends. "I'd known Charlie for a long time, but I'd never really worked with him before," Eggers recalled. Except in one area: "Charlie and I used to dry apricots together. He was great at drying apricots." Now Eggers was about to discover an untapped side of his old friend. "Charlie was a very serious, fine engineer, but up to that point, as I recollect, his latent outstanding leadership capabilities had not really shown up." They were about to come into full flower.

At the meeting, Eggers laid out his ideas for an Ames solar probe, to enthusiastic response. The next step would be a formal study, setting up a task group to investigate the possibilities and to lay out plans for presentation to higher NASA authority. Out of the blue

Charlie Hall found himself elected. "Al turns to me and says, 'Charlie, would you take care of that?'" Hall laughingly recalled in a 1999 interview. "That was typical Al. He didn't like details. He was a thinker and a policy maker, but don't ask him to do details."

Hall threw himself into the task with extraordinary gusto and energy, pulling together people not just from the VED but also from other Ames branches. "Charlie did an exquisite job of heading up that task group," said Al Eggers. The study was finished in 1961, and Hall began to present the results to various movers and shakers in the NASA and space science community. When Hall took his Solar Probe Committee to a meeting of NASA's Particles and Fields science committee at Stanford University, just down the road from Ames, he began to get excited. Several of NASA's top space science figures, including Chuck Sonett, were present at the meeting and encouraged Hall to pursue the project further. The enthusiastic reception Hall and his team received convinced him that they really had something—something more than just another intriguing feasibility study to be passed upstairs, dutifully nodded over, and filed into oblivion. He began to push for Ames to actually take on the project and build the spacecraft itself: "Eggers called me in one day after a few more presentations and said, 'What do you think we oughta do next, Charlie, update the report?' And I said, 'No, Al, I think we oughta go to headquarters and get some money and build the damn thing!'" Hall would not be stopped, and Eggers set him loose. "Charlie just grabbed that thing. He was going hard. I was delighted. I figured, all right, let's keep going. He put the vast majority of his time on it. He just grabbed the ball and ran."

It was time to put up or shut up, as Hall would so directly phrase it. That meant taking the show on the road to NASA headquarters, where the final decisions were made and the real money was handed out. "Looking back, I can't believe I was so brash in those days," Hall laughed. "Most of the people you make presentations to can't even spell 'space' yet, they're so young." After another presentation to another NASA science committee at headquarters, one official, Jesse

Mitchell, asked Hall whom he'd like to see at headquarters in order to plead his case. Being brash and deciding to go right to the top, Hall suggested Hugh Dryden, NASA deputy administrator. He was rather pleasantly surprised when Mitchell took the suggestion seriously. Unfortunately, Dryden's calendar was too full that particular day for such a last-minute appointment, so Mitchell suggested that Hall talk to Edgar Cortright, deputy administrator of space sciences.

Hall found Cortright unexpectedly receptive. Unknown to Hall, Cortright had been looking for a way to give some competition to the Jet Propulsion Laboratory in Pasadena, which since NASA's inception had been the prime center for unmanned lunar and interplanetary space projects. JPL's relationship with NASA had always been problematic and contentious. Officially, JPL wasn't run by NASA but by the California Institute of Technology, its parent organization, under contract *for* NASA. Under its headstrong director, William Pickering, JPL had put America's earliest satellites into orbit, and those achievements had made JPL both aggravatingly independent in outlook and rather smug about its own superiority. Those qualities didn't sit well with NASA headquarters, which tended to consider Pickering and JPL as a collection of spoiled-rotten scientific prima donnas. When Hall walked into Cortright's office, JPL was in the midst of a crisis with its series of Ranger lunar probes, each one failing more spectacularly than the last, to public dismay and congressional outcry. So if another NASA center was willing to put itself out on a limb and try to steal a little of JPL's unmanned thunder, Cortright had no objection, although Ames was hardly the place he would have looked for a contender for JPL's throne. He told Hall so, saying that Ames had the reputation of a college campus. Hall remembered Cortright telling him that "I think you're biting off more than you can chew with a solar probe" and instead suggesting that Ames consider a more modest program of interplanetary probes. Hall lacked the authority to commit Ames formally but agreed to raise the idea with Smith DeFrance and Al Eggers back in California.

"At headquarters we were interested but wary," Oran Nicks, then head of NASA Lunar and Planetary Programs, recalled in his memoir, *Far Travelers.* "While the project could fulfill a basic scientific need, and the Ames engineers had distinguished themselves in research activities, none of them had obviously relevant project management experience. The proposed project effort would clearly not be simple; one wondered how Ames, starting from scratch, would deal with the launch vehicle interface problems, the scientific community, and the challenging data acquisition problems that would have to be solved." At Ames, Hall found DeFrance completely open to the idea, something that surprised many of the old hands. As Eggers explained, "DeFrance had a long history of the old NACA way of doing business. Taking on major project responsibility was a whole new challenge." Yet DeFrance was also a shrewd administrator and undoubtedly realized that here was a chance to keep Ames from being left behind in the rush to space, a victim of its own past and unwillingness to change.

Over the next several months into 1962, Chuck Sonett, Oran Nicks, and other mavens of NASA's space sciences branch visited Ames. Meanwhile, Hall engaged Space Technology Laboratories, soon to rename itself TRW, for a feasibility study of an interplanetary probe. "We were now in a mode of finding common ground with NASA [headquarters] and at the same time finding out what kind of space probe might be most attractive from their point of view as well as ours," Eggers said. "It began to veer away from the solar probe idea towards what evolved into Pioneer."

That final concept was audacious in its originality and daring: a series of five solar-orbiting probes, each stationed at a different point either just inside or just outside Earth's own orbit around the Sun. No one had yet put spacecraft in such an orbit, but it would be the perfect place from which to monitor solar activity and to explore the nature of the interplanetary medium, thus collecting data that would be enormously valuable for any future manned or unmanned missions. The five spacecraft would be small, inexpensive, and nearly identical in design: spinning cylinders powered by solar cells covering the surface. And by stationing several spacecraft at different places in nearly the same orbit, simultaneous readings could be taken from widely separated points in space, something that had never been done before. They would become a network of deep-space observation posts covering the entire orbit of the Earth.

Eggers's original idea of a solar probe had metamorphosed into something quite different but still an endeavor that would move Ames into a more prominent role in America's space effort, as he'd originally envisioned. All that remained was to secure that elusive formal approval, and the funding that would follow, from NASA headquarters. With the support of the NASA space science community and the TRW technical study in his pocket, Hall was set for the final assault on the powers that be. He made another trip to Washington with Chuck Sonett, who had previously left headquarters and transferred to Ames to work on developing the project. This time, however, Hall had yet another weapon. Ames director Smith DeFrance had also traveled to Washington to attend the presentation and show his support. Normally, this might have been hardly worthy of notice. So a center director traveled to headquarters to fight for a project; big deal. In DeFrance's case, though, it raised more than a few eyebrows because DeFrance was infamous for never flying anywhere. After an airplane crash in which he'd lost an eye during his test pilot days at Langley, DeFrance had solemnly promised his wife that he'd never climb aboard another airplane. With a 4-day transcontinental train trip between his office at Ames and Washington, DeFrance was hardly a frequent sight in the corridors of NASA headquarters. So if he *were* there, there had to be a very good reason, and everyone knew it.

Hall and Sonett made their presentation to Robert Seamans and other key NASA brass, with Hall presenting the technical aspects of the mission and the spacecraft and Sonett arguing for the scientific rationale. After all of the studies, presentations, committees, and meetings, starting with Eggers's first gathering of his engineers, it was now up to Seamans and the NASA hierarchy to decide. Hall and Sonett felt good about their performance and the way they had melded together to give an effective presentation, but they knew only too well that this wasn't a talent show. Although a bad presentation could sink you, even a perfect one didn't guarantee anything. Competition was fierce for every slice of the NASA budget, especially for whatever was left over after the lion's share of funding went to the more visible and glamorous manned program. Hall knew he had a solid proposal, a good project, and good people; he also knew that Ames had never managed a major project before, and many still doubted that it could.

Almost 40 years later, Hall still vividly remembered what happened next. "Seamans turned to Smitty [Smith DeFrance] and said, 'Smitty, what do you think of this?' And my heart just dropped. I thought, God, he could kill it right now, do anything he wanted with it." Even Hall, at that point, wasn't fully certain of DeFrance's unequivocal support. Would DeFrance, the old NACA engineer famous for his traditional ways, put his beloved Ames at risk? He did: "He said, 'Ames is 100 percent behind it,'" Hall recalled. "And I knew we were going to get the program because DeFrance was extremely admired and well thought of at headquarters. And they knew he would be backing me in any way, shape, and form and wouldn't let the thing fail."

Hall was right. In June 1962, NASA headquarters informally approved the Ames Research Center to carry out the Pioneer project of interplanetary probes; official approval followed in November. The spacecraft would be numbered *Pioneer 6* through *10*, following in the footsteps of several of America's earliest spacecraft, although the new Pioneers would have nothing in common with their ancestors but the name. Ames would have its shot at project management and its chance to compete with the 500-pound gorilla of the Jet Propulsion Laboratory in the unmanned spaceflight arena. Charlie Hall was named Pioneer project manager. His work was just beginning. At the time he didn't realize that he'd just accepted what would turn out to be a lifetime position.

2

Reaching into the Void

Even while Charlie Hall was still working in the Ames wind tunnels and before his name became synonymous with the Pioneer project, the original spacecraft to bear the Pioneer name were ushering in the space age.

They were, for the most part, glorious failures. Before the birth of NASA in October 1958 placed the space program firmly in civilian hands (at least officially speaking), the earliest American space efforts grew out of competing military programs. Wernher von Braun and his fellow German rocketeers were the most prominent, working for the Army Ballistic Missile Agency, and they had saved face for the nation after America's first attempt to launch a satellite, the Navy's *Vanguard*, blew up in a humiliating fireball in full view of the world in December 1957. But if the Army was on top of the heap, the Navy hadn't given up, and the Air Force was nipping at both their heels.

With the Russians orbiting ever larger and heavier satellites and knowing that the Moon would be the Soviets' next target, the military decided to shoot for the Moon. The very first Pioneers were lunar probes funded by the Advanced Research Projects Agency of the Department of Defense. There would be five Moon shots, three from the Air Force and two from the Army. The first one, launched by the Air Force on August 17, 1958, never made it into space. The first stage of its Thor booster blew up 77 seconds after launch. To mark its igno-minious failure, this one was dubbed *Pioneer 0.* The next Pioneer, although also conducted by the Air Force, was actually the very first NASA space mission, launched less than 2 weeks after NASA's Octo-ber 1, 1958, birth. It was somewhat more successful—at least, it made it out of the Earth's atmosphere, although without enough velocity to escape the Earth's gravity and travel to the Moon thanks to an early shutdown of its second-stage booster. Still, the craft provided useful data on the Van Allen radiation belts and the fringes of interplanetary space before it reentered the atmosphere and burned up 2 days later. The prescribed mission had again failed, but to a nation anxiously grasping for any space laurels, it was enough for the moment. Al-though the Russians predictably dismissed the modest achievement, the spacecraft had made America's farthest penetration yet into space, thus earning the distinction of being forever known to history as *Pio-neer 1.*

The third Air Force moonshot, *Pioneer 2,* launched on November 8, 1958, would not enjoy even that much success. First its second stage burned out prematurely, then the third stage simply refused to fire. The craft made it to an altitude of just under a thousand miles before falling back to Earth and burning up, although it did manage to pass along some interesting data on micrometeoroids and Earth's equato-rial regions before its untimely death.

Von Braun and the Army came next. *Pioneer 3* was launched a month later, but again booster problems doomed the mission. Like *Pioneer 1,* it just wasn't given enough of a push to escape Earth's grav-ity and after reaching slightly over 63,000 miles, fell back homeward. Yet *Pioneer 3* confirmed the existence of the recently discovered Van Allen radiation belt and discovered a second radiation belt at a higher altitude. America hadn't reached the Moon yet, but it was getting closer, and learning some interesting things in the process.

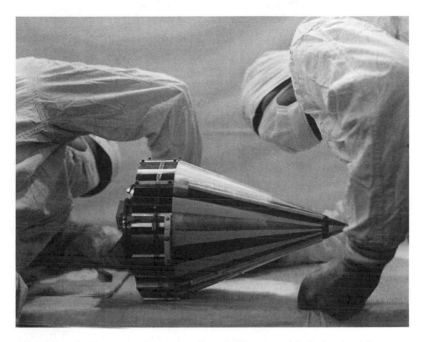

Technicians in "clean room" garb inspect the *Pioneer 3* probe before launch.

The next military Pioneer shot was von Braun's *Pioneer 4* in March 1959, and of all five this one came closest to success. It missed the Moon by about 37,000 miles but became the first American spacecraft to escape Earth's gravity. It then achieved solar orbit, sending back radiation data on deep space. Unfortunately, it was too little too late, at least as far as the space race was concerned. Two months earlier the Soviets had opened the new year with *Luna 1*: the first human spacecraft to reach escape velocity and travel to the Moon. After flying past the Moon, *Luna 1* became the first spacecraft to enter solar orbit. Despite America's best efforts, the Russians had won again. After *Luna 3* gave humankind its first look at the far side of the Moon in October 1959, NASA tried to send three more small spacecraft to the Moon over the following year to trump the Soviets, spacecraft that would have also borne the Pioneer name had they made it into space.

Two blew up in midair shortly after launch; the third suffered yet another booster failure and burned up in the atmosphere. For now at least, the Moon belonged to the Russians.

The next spacecraft to be called Pioneer, however, would do far more than go out in a blaze of failed glory across the sky. On March 11, 1960, *Pioneer 5* (or *Pioneer V*, using the system of Roman numerals in vogue at the time for spacecraft designations) became the first spacecraft to be intentionally launched into a solar orbit. "U.S. ROCKET PUT INTO SUN ORBIT—WILL BE FIRST TO GATHER DATA DEEP IN INTERPLANETARY SPACE," proclaimed the *New York Times* in its top headline of March 12.

After what seemed to be an endless parade of spectacular Russian successes and miserable American failures, NASA finally had something to crow about. The *Times* made no bones about this fact: "Within the space agency considerable psychological and technical importance was attached to today's long-postponed launching." In fact, NASA hadn't had a successful launch since the *Explorer VII* satellite 6 months earlier. "The agency was growing anxious for a success to give some much-needed prestige to the space program," the *Times* helpfully noted. At Cape Canaveral, "missile workers . . . rejoiced today as *Pioneer V* went roaring off toward an eternal orbit of the sun. 'That should shake up the rust,' the test conductor, Earl Wollam, remarked to his boss." At this early stage of the game, even the press hadn't yet settled on the proper nomenclature for space activities: the *Times* coverage repeatedly refers to *Pioneer 5* as an "artificial planet" or "planetoid," rather than the terms "probe" or "spacecraft" which would later become common coin.

The craft receiving all this attention was a 95-pound aluminum sphere built by Space Technology Laboratories and NASA's Goddard Space Flight Center, with four awkwardly angled solar cell panels protruding from its circumference like the paddles of a riverboat wheel, and carrying a 150-watt transmitter, the most powerful sent into space at the time. *Pioneer V* was originally intended for a Venus flyby, but after various delays in readying its Thor-Able rocket booster and the scientific payload, Venus had obstinately moved too far away in its orbit from Earth to allow mission planners much confidence in sending the craft there. With the flexibility born of both practical necessity and the burning desire to keep up with the pesky Russians, the spacecraft's objectives were changed to a solar-orbiting mission.

It didn't matter much to the press and the public, who saw *Pioneer V* as an unmitigated American triumph. For a country whose rockets always seemed to blow up while those of its chief competitor did spectacular things like photograph the far side of the Moon, sending a spacecraft (or "artificial planet") farther from Earth than anyone else was a big deal. The *New York Times* found deep philosophical meaning in *Pioneer V*'s voyage. "All men are brothers as they contemplate the mysteries, dangers and challenges of the vast realm into which *Pioneer V* is now penetrating," wrote the editors. "Differences of nationality, of religion, of merely terrestrial history pale into insignificance as we contemplate the vast international cooperation which will be needed to conquer this new dimension of human activity. . . . The true meaning of *Pioneer V* is that we have entered a new era in which the old nationalisms and petty rivalries among groups of human beings must give way to the need for joint action and generous cooperation which alone can meet the challenges of the future, in space and on this tiny planet both."

However arguable its cosmic significance or contributions to world peace might have been, *Pioneer V* provided some tantalizing data that whetted the appetites of space scientists worldwide. Not long after launch, it confirmed the existence of a "current ring" circling Earth in the outer Van Allen radiation belt. Less than 2 weeks later on March 31, more than 1.6 million miles from home, the sphere passed through the shock wave of a massive solar storm, providing the very first space weather report 4 to 8 hours before the storm hit Earth, sending compass needles swinging and disrupting communications. Obviously, interplanetary space was a much busier and more violent place than had previously been suspected, a realization that would have profound implications for future exploration—both unmanned and manned. Newspapers and wire services around the world closely followed the odyssey of *Pioneer V* until June 26, 1960, when at 22 million miles from home, it grew far too distant even for Sir Bernard Lovell's enormous radiotelescope at Jodrell Bank, England, to detect. It was the greatest distance anyone had yet communicated with a spacecraft.

But *Pioneer V* had done its job and more besides. It had proven that it was possible not only to send a spacecraft deep into interplanetary space but also to communicate with it, control it, and obtain valuable data from it. For 1960 and particularly for the United States, these successes were not to be ignored. And *Pioneer V*'s triumph was made all the sweeter by the string of failed Soviet missions to the Moon and to Mars that filled the remainder of the year. Although *Pioneer V*'s scientific achievements were soon to be overshadowed by other spacecraft, including the later Pioneers, it earned its place in history as a trailblazing feat of engineering and deep-space communications that provided the foundation for every deep-space mission, unmanned and manned, that followed it into the void. The fact that in the process it also helped to restore some of America's national pride and technological self-esteem was not an inconsequential bonus.

MAKING IT WORK

Aside from their common name, none of the early Pioneers had much in common, except, perhaps, their propensity for embarrassing failures. In the earliest days of spaceflight, the naming of American spacecraft was almost an afterthought, done mostly for the benefit of the press and the public. The military and governmental entities that launched space shots before NASA tended to favor the boring technological nomenclature of engineers' and manufacturers' designations: Able this, Baker that, Atlas-1B, Thor-Able C-3, and so on. Any more colorful and evocative monikers were more or less arbitrary, hardly indicative of any sort of single central authority operating and launching the craft or of any common design or mission objectives. One example is the series of myriad Explorer satellites that followed America's first satellite, *Explorer 1*. A more bewildering and varied array of spacecraft would be hard to imagine.

By 1962 NASA was having none of this sort of confusion anymore. Each project would now have a well-defined and specific name that would apply to that individual project and none other. Perhaps inspired by the triumph of *Pioneer V* that redeemed the name by erasing the dismal memory of the exploding boosters and burned-up capsules of the initial Pioneers, NASA decided to revive the name for the new series of interplanetary probes to be managed by Ames Research Center and Charlie Hall. From now on, no other American spacecraft would bear the name of Pioneer, save those constructed and operated under the aegis of the Ames project. As far as official history was concerned, the earlier Pioneers just happened to have the same name; although the numbering of the Ames Pioneers would begin with the number 6 to (reluctantly) acknowledge the existence of the earlier craft, the first official Pioneer mission under Ames would be designated *Pioneer A*.

Whatever the project was called, with its official approval by NASA headquarters, Charlie Hall now had his hands full. He loved it. "He was a transformed man," Al Eggers recalled. At first, however, Hall discovered that the enthusiasm of some people for a hypothetical study didn't necessarily translate into real-life commitment. "I figured that when we got the go-ahead for the project, people at the lab would beat down the doors . . . by God, I didn't hear from anyone!" It was the old NACA Ames culture hanging on yet again: the reluctance to do anything but pure research, avoiding the responsibilities and headaches of a major project. Describing the dilemma, Hall echoed Ed Cortright's words: "Ames in those days had a very university atmosphere; they didn't want to have their research sullied by the outside world."

With the help of other Ames people, Hall bulled forward, gathering his forces and a core group of people to get the project off the ground, hiring some from outside Ames. One of the first things he needed was office space, always at a premium at Ames. Hall solved the problem in characteristic fashion. When formal offices were slow in being assigned to his group, he set up cubicles in the Ames cafeteria and before too long ended up taking over the entire space. "Another place had to be built where people could go eat once in a while," Eggers remembered, laughing.

But Hall had bigger things to worry about than lunch. With Pioneer approved and people to staff it, he now had to find a prime contractor to actually design and build the spacecraft. The Pioneer office put out a request for proposals to various companies, and after initial evaluation the field was narrowed to four candidates: TRW (formerly Space Technology Laboratories, the builder of the earlier Pioneers), Westinghouse, Bendix, and Hughes Aircraft. To make the final selection, Hall and some of his staff embarked on a cross-country journey to visit each of the competitors in turn. It was to be a very educational trip and one that was vital in the development of Hall's managerial style. "That taught me one thing," he remembered. "You can read these proposals and boy, they sound great. You think they have everything in the world." But Hall found that the glowing promises of a formal proposal and the reality of a company's resources could be worlds apart. One potential Pioneer contractor had "beautiful equipment," but it was all brand-new and untested, and none of the engineers present even knew how it worked. The company was completely inexperienced in space, looking for a foothold in the business, hoping that Pioneer would provide it. "It would have been a disaster if we'd picked them," Hall said.

After the fact-finding trip, the choice was narrowed down to two candidates, TRW and Hughes. There seemed to be compelling arguments for each. "Several of the technical aspects of TRW were better than Hughes, but Hughes had a lower price estimate," Hall explained. Following further analysis of both bids, Hall discovered that the gulf between the two cost estimates wasn't quite as glaring as it originally seemed. TRW and Hughes used very different accounting methods, which had resulted in an artificially low Hughes price. Another piece of Hall's managerial education had fallen into place: corporate accountants, and their cost estimates when bidding for a highly desirable contract, weren't created equal.

TRW won the Pioneer contract and set about designing the spacecraft as the Pioneer project office began to solicit noted scientists for experiment proposals. There was no time to waste. Plans called for launching the first Pioneer late in 1964, to coincide with the International Quiet Sun Year (IQSY), a worldwide effort to study the interplanetary environment during a predicted tranquil period in the solar cycle. With a relatively benign period of solar activity, scientists hoped to build up a picture of deep space without the interference of sunspots, solar storms, and flares that would flood the solar system with high-energy radiation and charged particles. The IQSY was modeled after the International Geophysical Year of 1957-1958, during which the first satellites discovered Earth's radiation belts. In effect, the IQSY was a "sequel" to the IGY for astronomers and geophysicists.

On January 30, 1963, the NASA Office of Space Sciences sent out an Announcement of Opportunity to scientists around the country. The AO provided basic information about the Pioneer mission and the expected characteristics of the spacecraft, not to mention the required NASA format for the proper submission of experiment proposals. It was enough to send planetary scientists from one coast to the other scrambling to get their proposals together by the April deadline. It wasn't much time in which to conceive an original experiment and design a preliminary instrument to carry it out, but space scientists were used to working under such constraints by now. Time was hardly as troublesome an issue as other factors: the instrument's weight, size, shape, and power requirements were always drastically restricted in space travel. Trying to meet all the inevitable limitations that would be imposed on your experiment, while still remaining within a reasonable budget that wouldn't alarm either NASA or your academic institution, was less a scientific endeavor than a rather peculiar art form. Some scientists, such as the University of Iowa's James Van Allen and the University of Chicago's John Simpson, were virtuosos of the art. But they had years of experience in dealing with the labyrinthine world of government contracts, dating back to long before NASA had even existed. Van Allen, for example, had placed his instruments on the captured German V-2 rockets and their American descendants, along with high-altitude balloons and other military-funded projects. Simpson had done similar work and had been a member of the Manhattan Project years before America began shooting rockets into the upper atmosphere. Such experience and expertise didn't make the challenges any less daunting but did make them somewhat easier to meet.

TRW's design didn't leave much wiggle room for the six experimenters finally chosen, officially known as principal investigators (PIs). The spacecraft was essentially a simple cylinder, 37 inches in diameter and 35 inches high and weighing only 140 pounds. Somehow, within that volume and that weight limit, everything had to fit: not only the scientific instruments that were the spacecraft's *raison d'etre*, but everything else that would allow it to operate, including power, communications, and data instruments. In an age when transistors and other electronic components were still discrete, individual physical objects instead of scratches on microscopic silicon wafers, the problem of miniaturizing such a complex piece of hardware strained the ingenuity of even the most talented engineers. The final spacecraft design would have over 56,000 parts, and that was while keeping it as simple and basic as possible.

Since a large part of its mission would be observing the effects of magnetic fields on charged particles, the Pioneer spacecraft had to generate as little magnetism as possible itself. Otherwise the observations of the magnetometers and charged particle instruments would be corrupted and their data rendered useless, a problem that had plagued many past spacecraft. "Magnetic cleanliness" was paramount. Fortunately the TRW engineers were up to the task. "Today it would

The *Pioneer* 6 through 9 solar orbiting spacecraft all shared this basic design.

be easy because we know how to do it," Hall said. "In those days it was something brand new. It just didn't have any magnetic field at all."

Creating a spacecraft within the required parameters was only part of the problem. This would not be a chunk of equipment intended to sit in the benign environment of a government workshop or university lab, where its worst enemies might be bored technicians or ham-fisted undergraduates. The Pioneer spacecraft had to operate in a place bathed in radiation, with extremes of temperature, and peppered by dust and micrometeoroids, and it had to do so millions of miles from Earth. No technician would be able to tweak it with a well-placed screwdriver if it started acting twitchy, and no repairs would be possible if something went wrong so far from home. The spacecraft couldn't be sent back to the factory for warranty service if it failed. Once launched, it would never see Earth again, and no human hands would ever again touch it. Pioneer not only had to work and do what it was supposed to do, it had to be *reliable*. One hundred percent reliability was an unattainable fantasy, of course, but anything much less would probably doom the mission.

Because it operates in an environment without any natural up, down, left, or right, every spacecraft needs some way to orient itself and maintain its bearings, so that it can do its job and keep in touch with Earth. One technique for achieving this is called three-axis stabilization, which uses small thrusters to adjust the spacecraft's orientation along its pitch, yaw, and roll axes. Although an efficient system, three-axis stabilization is also quite sophisticated and thus complex, expensive, and prone to malfunctions. To keep Pioneer simple and hence more reliable, its designers opted for the other method of spacecraft orientation: spin stabilization. The craft is simply set spinning along its long axis like a top. Like a spinning top, it is then inherently stable and tends to maintain its orientation. Unlike a top on Earth, a spacecraft doesn't have gravity or the friction of a tabletop to slow it down and stop it, and just keeps spinning by itself, without any further intervention required. It's an elegant and effective technique that's been used since the beginning of the space age. There are disadvantages, of course; the cameras and instruments of a constantly spinning spacecraft can't focus for more than an instant on a particular point of interest, and the design of the craft's communications antennas must account for their constant motion. But for an inexpensive solution to the spacecraft orientation question, spin stabilization is the only answer. It would prove to be one of the hallmarks of the Pioneer program under Charlie Hall.

TRW was confident in its design and its engineers, but the company was also realistic. The *Pioneer A* through *E* spacecraft would be built to last approximately 6 months to a year each, TRW managers told NASA. With the planned overlapping launch schedule for the five spacecraft over the next several years, that was enough to ensure that there would always be at least one, and for at least a brief time more than one, spacecraft operating at all times. There were no guarantees, of course; every interplanetary spacecraft is one of a kind, not a product of an assembly line. The cost-plus-incentives contract NASA gave TRW provided bonuses if the company built the spacecraft on schedule and within budget, but extended warranties for space probes didn't exist. Still, a minimum 6-month design lifetime was more than enough to satisfy the stated objectives of the program and in the early 1960s sounded like a perfectly reasonable life expectancy for a deep-space probe.

In what was rapidly developing into a rather infamous hands-on style, Charlie Hall made sure that TRW and its subcontractors stayed on the ball. Representatives from the Pioneer project office were on hand at TRW and wherever else project work was proceeding, ready to help solve problems, answer questions, and coordinate activities. Hall held frequent meetings and reviews to keep the project on track and head off major glitches before they occurred. He was already earning a reputation as a firm but fair taskmaster, one who seemed to have a knack for inspiring people to their maximum effort.

By December 1965, preparations were well under way for the launch of *Pioneer A* (which would earn the designation *Pioneer 6* upon a successful launch). The craft was at Cape Canaveral (now renamed Cape Kennedy by President Johnson), checked out, and mounted atop its Thor-Delta E booster rocket, ready to go, by the middle of December. There would be a slight delay, however.

Earlier in the month, astronauts Frank Borman and James Lovell had gone into Earth orbit aboard *Gemini 7*, and were still in space on a 2-week endurance mission. Their colleagues on *Gemini 6*, Wally Schirra and Tom Stafford, had been scheduled to launch back in October for a rendezvous with an unmanned vehicle but were still on the ground 2 months later due to a technical glitch that had blown up their rendezvous target after its launch. Instead, NASA officials decided to let Schirra and Stafford rendezvous with the still-orbiting Borman and Lovell, meaning that for the first time NASA would have two manned spacecraft in space simultaneously. That would strain

Pioneer 6 is mated to its Delta third stage. This assembly would later be placed atop the much larger Thor booster for launch.

the communications and tracking resources of NASA to their limits, and with the lives of four human beings in the balance, there were no resources to spare for the launch of an unmanned spacecraft. *Pioneer 6* would have to wait. After *Gemini 6* was safely launched on December 15, Pioneer followed it into space the next day. *Gemini 6* and *7* went on to make the historic first rendezvous in space, while *Pioneer 6* left Earth behind and took up its station in solar orbit, operating perfectly.

The Ames Research Center had proved to the doubters within NASA that it could plan, build, and manage a space project. Now it was the scientists' turn to prove the worth of their experiment proposals, and *Pioneer 6* didn't disappoint them. Spinning sedately at 60 revolutions per minute as it orbited the Sun, the spacecraft sent back reams of data on the solar wind, interplanetary radiation and micrometeoroids, magnetic fields, and cosmic rays. A fire at the Jet Propulsion Laboratory in May caused a brief loss of communications for several days, but *Pioneer 6* kept transmitting its discoveries. The Pioneer PIs quickly found themselves overwhelmed by the wealth of information, so much so that they became concerned about the coverage the spacecraft was receiving from NASA's Deep Space Network, which was already overloaded with the work of tracking other manned and unmanned craft. It was a dilemma that continues to plague space scientists to the present day.

Pioneer 6 passed its 6-month anniversary still operating flawlessly, and in August 1966 its identical twin *Pioneer 7* was launched, taking up a position slightly inside Earth's orbit around the Sun. Now the two spacecraft could begin to operate in tandem, and in October they detected the strongest solar flares yet witnessed. When not monitoring the heaving convolutions of the solar atmosphere, *Pioneers 6* and *7* continuously probed regions of interplanetary space that had never before been explored.

To the extreme pleasure of Charlie Hall, the PIs, and NASA headquarters, the Pioneer program was cruising along smoothly, with nary a hitch. *Pioneer C,* that is, *Pioneer 8,* followed its sister craft into solar orbit on December 13, 1967, as did *Pioneer D/9* on November 8, 1968. While carrying a somewhat different set of scientific instruments than their predecessors, *Pioneers 8* and *9* shared their basic design and mission, with *Pioneer 8* essentially sharing Earth's solar orbit and *Pioneer 9* orbiting slightly beyond.

Only one major mishap echoed back to the disastrous Pioneers at the beginning of the decade. The last of the solar-orbiting craft, *Pioneer E*, was launched on August 27, 1969 on its way to earn the name of *Pioneer 10*. "Ten seconds before the end of the first-stage burn, the rocket hydraulic pump went out and we lost control of the direction of the rocket," Hall recalled. "We have pictures of that thing turning around 270 degrees, and after the turn the second stage separated. So the second stage went off to the right and the first stage fell down into the Atlantic Ocean. They blew up the launch vehicle, but we tracked the spacecraft all the way into the Atlantic Ocean." The *Pioneer 10* name would be reserved for a future spacecraft.

Because all four of the earlier Pioneers were still humming along quite happily 3 years after the first of the quartet had been launched— against all expectations of NASA, TRW, or the scientists—Hall found himself running what was in effect Earth's first space weather network. The four Pioneers arrayed around the Sun formed a sort of interplanetary distant early-warning system for the detection of massive solar disturbances that could affect activities on Earth. Solar flares, sunspots, and storms pour out enormous amounts of radiation and magnetism that can overcome Earth's protective magnetic field and interfere with electric power grids, worldwide communications, aviation, satellites, and even weather on the ground. The Pioneer project office established a liaison with the National Oceanic and Atmospheric Administration, which passed along warnings of solar disturbances to other agencies such as the Federal Aviation Administration, the airlines, and power companies.

As the Apollo astronauts ventured beyond Earth to lunar orbit in 1968, and then to the surface of the Moon from 1969 to 1972, the Pioneer quartet also watched over them. Although lunar missions were scheduled for periods when solar activity and the accompanying lethal radiation it generates were at a low ebb, the prediction of such solar outbursts was hardly (and still isn't) an exact science: violent flareups are always a possibility at any time. When humans were away from Earth's protective environment, the Pioneers provided their first line of defense. Had any deadly solar flares occurred, NASA would have received advance warning, giving the astronauts precious time to protect themselves by altering their orbit or, if on the lunar surface, taking refuge inside their lunar module until the storm of radiation had passed. Such functions hadn't been one of Pioneer's original objectives, but the unexpected longevity of the spacecraft had yielded them as a bonus.

The surprising endurance of *Pioneers* 6 through 9 had a downside, however. NASA's worldwide Deep Space Network had only so many big radio dishes to go around and only so much capacity to handle the burgeoning number of unmanned and manned space missions of the 1960s. Communicating with Earth satellites was one thing; receiving data and sending commands to a craft millions of miles away was quite another. The largest antennas of the DSN were 210 feet (i.e., 64 meters) in diameter, and there were only three of them: one in the California desert at Goldstone; one near Madrid, Spain; and the third near Canberra, Australia. But with only 24 hours in a day, and each antenna capable of communicating with only one spacecraft at a time, it was impossible to stay in touch with all of the spacecraft beyond Earth all the time. DSN tracking time was a finite commodity, to be scheduled, prioritized, and doled out according to all sorts of factors. If, for example, DSN facilities were needed to support an Apollo mission where human lives were at stake, the priorities were obvious. If an unmanned planetary probe was approaching its destination and entering its encounter phase, the most crucial and central part of its mission, coverage for that spacecraft took precedence over others that were in a less critical phase of their missions. Competition for DSN coverage among different NASA projects was always intense, was always a major headache for any project manager, and was sometimes affected by concerns other than the purely technical or scientific.

As NASA's deep-space activities intensified into the late 1960s and the 1970s, the Pioneer PIs found themselves being shunted aside for more current missions, such as the Mariner flights to Mars and Venus. The primary mission of each Pioneer spacecraft officially ended about a year after its launch; after that the craft was considered to have fulfilled its mission objectives and to then enter its "extended mission," with lower priority than missions that were still active. The Pioneer PIs were lucky to get several hours of data per week, quite a letdown from the nearly continuous stream they had enjoyed earlier in the project. Some of the scientists, such as James Van Allen and John Simpson, began tirelessly lobbying NASA and Charlie Hall, arguing that the value of their scientific findings was being compromised by the gaps in their data caused by the sporadic DSN coverage for Pioneer.

They had a valid point. The processes and phenomena under study by *Pioneers* 6 through 9 were not only constantly changing, they required uninterrupted observation over time in order to ferret out their secrets. Only that would reveal the patterns and cycles of cosmic ray flux, the ebb and flow of charged particles and other radiation, and the differences at various regions around the Sun that might provide insight into the workings of the Sun and the interplanetary medium. Yet however valid their scientific rationales, Van Allen, Simpson, and the other experimenters were in direct conflict with other scientists on other projects whose technical concerns and scientific arguments were just as reasonable.

The extent of the problem, not to mention its perennial existence, is forlornly evident in the correspondence of the Pioneer scientists. Even as early as October 1966, only months after *Pioneer 7* had been launched, John Simpson fired off an aggrieved telegram to Charlie Hall: "I am shocked and disappointed at *Pioneer 7* coverage in view of the excellent scientific work being undertaken by the spacecraft. Present coverage fails completely to take advantage of this opportunity. . . . We cannot conduct our scientific research on such sketchy coverage." In a follow-up letter to NASA headquarters, Simpson complained: "I cannot perform experiments since I need continuous data for more than one solar rotation. If NASA cannot support programs and promises that are already made, there is little hope of proceeding to develop this area of space science."

Simpson's concerns were echoed by Norman Ness at the Goddard Spaceflight Center. "*Pioneer 7* status report dated November 8 indicates coverage on satellite becoming almost nonexistent," he wired to Hall and NASA headquarters. "Same is true of *Pioneer 6* although to a much larger degree. Request that the project make every attempt to alter situation and improve percentage of data coverage."

Unfortunately for the scientists, the problem was one that, while it might improve periodically, would never go away, even as the DSN continued to build new facilities and upgrade its capabilities. A late June 1969 memo to Simpson from one of his team members makes vividly clear the sorts of conflicts and compromises that were now an inescapable fact of a space scientist's life. "The Mariner encounter [the *Mariner 9* flyby of Mars] will affect the 210' antenna coverage. Calibrations for that experiment will start on 10 July and continue until approximately 18 July. During this period the overall Pioneer coverage will drop slightly. Beginning 30 July through 5 August the Pioneers will get no coverage; this period corresponds to the encounter. . . . The *Apollo 11* mission will affect the coverage of Pioneers during the period 18 July-24 July. There may be a chance for some coverage on Pioneer on a pot-luck basis. . . . During the period 10 August through 15 September the 210' antenna will be down for maintenance. As far as I was able to determine no coverage at all will be available to Pioneer during this period."

Perhaps one of the greatest ironies of the space age was that, at a time when NASA was embarking on its most ambitious adventures in both the manned and unmanned realms, the very scope and reach of those missions made it impossible for every individual project to achieve its ultimate potential—and that the unexpected success of a project such as Pioneer actually limited its ability to expand on that very success. As Alois Schardt, one of the NASA headquarters people whom Simpson importuned on Pioneer's behalf, remarked, "It is regrettable that we are unable to better capitalize on our good fortune in achieving longer spacecraft lifetimes."

Schardt wrote those words to Simpson in 1967. They would turn out to be remarkably prophetic.

CHECKING IN

At about 4 P.M. Pacific Time on December 8, 2000, Larry Lasher was sitting in front of a Macintosh computer in the Space Projects Facility building at the NASA Ames Research Center in California, waiting. He had been Pioneer's final project manager when the cancellation boom came down in 1997. He still operated in that capacity unofficially, though it was a job that hardly took much time away from his more immediate projects, most of them involving the study of planetary rings by spacecraft such as Cassini and Galileo.

The building was at the corner of Hall Road and Pioneer Avenue, although those weren't the streets' original names; Ames had recently renamed some of its roads as part of its 60th anniversary celebration in 1999. The room in which Lasher sat was utilitarian, functional; some parts were walled off by glass panels to separate them from the rest of the room. Computer equipment, some brand new, some of it painfully obsolete, lined the walls and nestled on desks, some of it functioning, some of it partly disassembled and in the process of repairs or upgrades. There were even a couple of computer systems in the room that wouldn't have been out of place in a museum, except that they were still needed for some very specific tasks that the smaller, sleeker desktop systems scattered about couldn't handle.

Lasher wasn't alone. Dave Lozier and Ric Campo were nearby, also bent over computer screens. All of the men had other formal jobs at Ames. Their presence in this room, the Pioneer control center, was semiofficial at best, because officially the Pioneer project no longer existed, at least as far as NASA was concerned. But while the stroke of a headquarters bureaucrat's pen on a piece of paper might formally terminate a project administratively and send people to other jobs or the unemployment line, it couldn't erase the spacecraft that had been built and sent far from Earth into the interplanetary void. It existed and would continue to exist whether stubborn humans back home chose to admit it or not.

Unfortunately, although the spacecraft might still exist and even continue to function, the only way to find out was to try to contact it. And for that you needed facilities, people, and time, and to secure those you needed the forbearance of the bureaucrats who called the shots. Usually, getting such consideration was next to impossible, especially when budgets were so tight and resources spread so thinly. But Lasher was a persuasive type and had managed to wangle special permission this time. Besides, even the beancounters realized this was something of a special occasion. Earth was about to talk to a faithful long-time servant that was far from home.

The signals started to trickle in from the Deep Space Network and show up on the Macintosh screen. The DSN had locked on, and data was coming in. It would continue to do so for about 2 1/2 more hours. The data rate was only 16 bits per second, ridiculously slow by year 2000 standards, but in this case the bit rate didn't matter so much as the fact that it was coming in at all.

Lasher, Lozier, Campo, and the others in the Pioneer control room were all smiles. A few people applauded. They had reestablished contact with humanity's oldest functioning spacecraft. For Dave Lozier, who had joined the project in 1966 and had been working with Pioneer longer than anyone else in the room, the moment was especially sweet.

Thirty-five years after it had left the planet of its origin, 83 million miles away, *Pioneer 6* was still alive, still working, and calling home.

3

Something Man Has
Never Done Before

The space beyond Mars was the great unknown. Even by 1969, 12 years after the first rudimentary spacecraft began probing outer space, and even as humans were walking on the Moon, they had yet to reach farther out into the solar system than the fourth planet from the Sun. Spacecraft such as *Pioneers 6* through *9* had successfully surveyed as far away from Earth as the opposite side of the Sun, about 180 million miles away. The two planets closest to Earth, Venus and Mars, had seen several American and Russian spacecraft pass by, sometimes still working, sometimes not.

The successes were most definitely few and far between. After *Mariner 1* was destroyed shortly after launch, NASA's *Mariner 2* made it to Venus in December 1962, becoming the first spacecraft to successfully fly by another planet and at last scoring one notable first for America, after several failed Soviet attempts at Venus. *Mariner 4* made the first successful flyby of Mars in November 1964. It took the Russians until 1967 to accomplish a fully successful Venus mission with *Venera 4*, after 13 failed attempts. Despite multiple tries, the USSR didn't make it to Mars until 1971. Whatever criticisms one might have about Soviet politics and society, even the most fervent anti-Marxist had to give the Russians an E for effort.

The few victories were preceded, and followed, by seemingly endless and heartbreaking near-misses and outright failures on both sides. Sometimes spacecraft never made it into space, destroyed by exploding boosters after launch. Others got into space but ended up trapped in Earth orbit. Still other craft were placed flawlessly on course for their destination but then suffered mysterious communications failures. Sometimes a probe lost its orientation and ended up spinning drunkenly out of control, becoming just another piece of whirling space junk hurtling into the solar system. Worst of all were those missions that seemed to go perfectly until just before the spacecraft arrived at its target, leaving disappointed scientists shaking their heads and aggravated administrators gnashing their teeth back on Earth.

Obviously, sending craft on a precise trajectory to rendezvous with another planet wasn't as straightforward a proposition as the mathematics and engineering plans might make it seem. The few successes at Venus and Mars had been tantalizing and had whetted scientific appetites for further missions to those worlds, but planetary scientists didn't want to stop there. They were itching to get closer looks at the other planets, particularly the gas giants Jupiter and Saturn, worlds that were radically different from the planets of the inner solar system for reasons that were barely understood. But the notion of sending space probes into the outer solar system was daunting at best and considered outright ludicrous in some quarters. Just getting to Mars had turned out to be challenging enough.

The series of failed missions to Mars inspired one journalist, Donald Neff of *Time* magazine, to half-seriously posit the existence of an evil "Great Galactic Ghoul" somewhere out around the orbit of Mars that liked to destroy, eat, or just toy with any spacecraft from Earth that dared to trespass into its realm. It was as good an explanation as any, some thought, including some engineers who hung a caricature of the imaginary beast on a wall at the Jet Propulsion Laboratory.

Great Galactic Ghoul or not, there were real concerns about the prospects of any spacecraft dispatched so deeply into space. Beyond Mars and before Jupiter lies the Asteroid Belt, a vast gulf in which millions of rocks orbit the Sun. The asteroids range in size from hundredths of an inch to hundreds of miles in diameter. For a spacecraft moving at thousands of miles an hour through this region, an impact with even a tiny fragment could cause serious damage; a bigger rock would be instantly fatal. The problem was that no one knew just how densely the Asteroid Belt was strewn with its rocky inhabitants. Scientists were reasonably sure that it would be possible to maneuver a spacecraft around the truly large asteroids, many of which were already in well-known and predictable orbits. But how much additional junk drifted between the big chunks? Would a transiting spacecraft be peppered by tiny particles like a blast from a shotgun or was the distribution of material sparse enough that there was little or no risk?

Even if the Asteroid Belt was survivable, it was only the beginning of a litany of problems that had to be faced in an outer solar system mission. The much greater distance between Earth and, say, Jupiter, from well over 300 million miles to as much as 600 million, compared to a relatively short jaunt of some 26 million miles to a place like Venus, meant that travel times had to be measured not in months but in years. So any spacecraft sent on such a voyage had to be sturdy and reliable enough to last the trip. It would do nobody any good to send a craft on a 2-year odyssey to Jupiter when it probably wouldn't live longer than a year.

Distance figured into the other major headaches, including how to power the spacecraft. Solar cells and panels proved effective for the earlier Pioneers, not to mention the Mariners and Explorers and Surveyors and all of the other craft that operated no farther from the Sun than the Earth or still closer. Even a trip to Mars was quite possible using solar power. But the intensity of the Sun's light at Jupiter is only 1/25 as strong as at Earth, which meant that a solar panel would have to be at least 25 times bigger in area to provide the same amount of power as it would at Earth. Larger solar panels would be expensive, and more than that, they would be terribly fragile, prone to being damaged by micrometeorites, being sheared off by the movement of the spacecraft or failing to deploy in the first place because of mechanical difficulties. Batteries were also a completely impractical solution: the length of the mission would require so many of them that the craft would be too heavy to launch.

Without adequate communications with a spacecraft, nobody was going to see any pretty pictures or receive any fascinating data. But communications are limited by the speed of light, 186,282 miles per second—a limit practically unnoticeable on Earth but extremely obvious at hundred-million-mile distances. The delay between sending a command to a spacecraft and receiving a response would be measured in hours, which meant that the craft would be effectively on its own much of the time. And a deep solar system space probe not only had to be tracked over such immensities, but its signals had to be picked out amidst all the other radio noise of space, natural and humanmade.

Finally, the very unknowns that such a spacecraft would be sent to reveal were a threat. Scientists knew that Jupiter was surrounded by intense bands of radiation. But what kind of radiation and, more important, how intense? Strong enough to fry the circuits of any hapless spacecraft that happened to wander into its clutches? If so, how close could a probe get to the planet and still survive? Close enough to perform some valuable science, or was the environment so lethal that there was no point in even making the attempt to approach the planet?

All of these vexing safety questions were matched by equally haunting scientific ones. All the data on Jupiter, Saturn, Uranus, and Neptune had come from Earth-based observations with optical and radio telescopes. These data had told scientists a lot but left much unexplained and raised even more questions. Jupiter, an immense ball consisting mostly of hydrogen, was known as the largest planet in the solar system, apparently gaseous or liquid without any true solid surface. Yet it seemed to emit more energy than it took in from the Sun. Why? What caused the radio transmissions that emanated from the planet? What were its moons like, and just how many of them were there? What was the Great Red Spot, so large that even early astronomers equipped with the crudest of telescopes noted that it engulfed the bands of colored clouds across the planet's face? Why were Jupiter and its outer planetary companions so different from the more solid, rocky inner planets such as Earth?

Scientists had to get a closer look to find out. Sending humans, of course, was out of the question: too expensive, too dangerous, too politically risky. Moreover, it was unnecessary. Unmanned spacecraft could do the job for much less money and much less trouble. But while few scientists needed to be convinced of the importance of space exploration, talking NASA and Congress, two agencies heavily invested in the political realities of the manned space effort, into fronting the money for an unmanned pure science mission was another thing entirely. Particularly after John F. Kennedy's goal of landing a man on the Moon and returning him safely to Earth had been achieved in July 1969, the people who controlled the purse strings saw less and less reason to spend money on space. There was a war on in Southeast Asia, after all, and with social unrest plaguing the campuses and poverty consuming the inner cities, surely the planets could wait. Was it worth spending millions of dollars to send probes into deep space just so that some scientists could gain a better understanding of Jupiter's particles and fields?

The answer, of course, was what it had always been. The search for knowledge was a noble human aspiration and worth pursuing for its own sake. Such arguments didn't sway the critics, but others did, such as those put forth by space advocates who demonstrated the actual fraction of the total NASA budget that unmanned missions consumed was minuscule and how NASA's budget itself had been rapidly shrinking as the Apollo program waned. And there was always the pesky Soviet Union. America had won the race to the Moon hands down, but the Russians had a nasty habit of surprising us. A successful mission to the outer planets would be a major coup, even without astronauts onboard.

And there was another consideration: maybe the planets *couldn't* wait. Back in 1961, a UCLA mathematics graduate student named Michael Minovitch working at the Jet Propulsion Laboratory (JPL) developed a method of propelling a space probe by using the gravitational field of a planet to boost it on course to another planet, the first planet acting as a sort of slingshot. A few years later a Caltech graduate student, Gary Flandro, expanded on Minovitch's work. In a 1966 paper, Flandro pointed out that direct missions to the outer planets "are characterized by high launch energy and very long flight duration. At least the latter of these two factors must be reduced if practical exploration of the outer solar system is to be accomplished. A very attractive source of energy which can be tapped to bring about this reduction is the gravitational perturbation of an intermediate planet."

Flandro also noted that the late 1970s "abounds in interesting multiple-planet missions utilizing massive gravitational perturbations of Jupiter." In other words, for a relatively short and precious period, the immense gravity of the largest world in the solar system could provide a free ticket to all the planets beyond it. The technique became known as gravity assist, and it held remarkable promise. Jupiter, Saturn, Uranus, and Neptune were inexorably drawing together in their stately orbits around the Sun in an alignment that occurred only once every 175 years. By the late 1970s they would be positioned such that it would be possible for one spacecraft to visit all of them, using the gravity of one planet to swing its course around for the next. Instead of a series of individual missions to each world, nearly the entire outer solar system could be explored with a single probe on a single mission. What made such a mission feasible was a natural phenomenon, a happy accident of simple orbital mechanics, and it depended not at all on politics, budgets, or popularity contests. It was an unprecedented opportunity for exploration that scientists took to calling the "Grand Tour."

But the Grand Tour had to be launched no later than about 1977 or 1978. After that the outer planets would have drifted too far apart from each other to make the gravity assist technique feasible for the tour. Time was wasting and budgets were shrinking. With the end of the Apollo program in sight, politicians were looking to shift America's emphasis from exploring space to *exploiting* it: making the space effort "pay for itself." The president and various congressmen might still wax poetic about building lunar bases and sending fabulous manned missions to Mars, but in reality they were directing money into the new idea of a reusable space shuttle instead. Supposedly, a space shuttle would make space practical, transforming it from an esoteric playground for egghead scientists and eccentric engineers into a new frontier for American business, technology, and—although this part wasn't touted too loudly to the public—the military, which was one of the shuttle's major supporters. Politicians liked to pay lip service to the ideals of pure science and exploring the universe, but they rarely put their money where their mouths were.

Even if the Grand Tour became a reality, though, it was still far in the future. Meanwhile, ideas for a mission to Jupiter had been kicking around NASA and the scientific community for some time. As early as February 1965, at an American Astronomical Society symposium in Denver, Colorado, on unmanned exploration of the solar system, a pair of engineers from TRW presented a paper with some fascinating ideas. Although they didn't realize it at the time, they were setting out the basic parameters of Charlie Hall's next job.

"The progress made by spacecraft exploration since the first Earth satellite in 1957 has been remarkable," said Herb Lassen and Robert Park. "Still, the fact that missions to the planet Jupiter and beyond are presently under serious consideration appears at first glance to be almost incredible." With the irrepressible optimism of the creative engineer, Lassen and Park continued: "However, the difference between a mission say, to Mars, and one to Jupiter is essentially no more than having a slightly larger booster and satisfying the increased lifetime requirements." Optimistic, yes, but also practical: "Of course, solar power will not be suitable at this distance from the Sun and thus another power source must be provided."

Lassen and Park also considered the big picture in a few more remarkably prescient comments. "The function of the first mission is to establish boundaries within which subsequent missions can make more accurate measurements. Therefore, we assume that the exploration should be carried out in two steps: the first step is to use a small, simple precursor spacecraft whose function is to establish the boundaries. Once the data from the first-generation spacecraft have been returned to Earth, it will be possible to design a more sophisticated system." Some of their words could have been written by Charlie Hall himself. "The first ingredient in designing a system for such long lifetimes is that the system must be inherently reliable—that is to say, that the system must be as simple as possible and must be adaptable to an evolutionary growth in reliability. . . . Perhaps the most important element in achieving the high probability of mission success is to establish a set of objectives which are reasonable—that is, within the state of the art of the spacecraft and boosters."

In the interest of keeping things simple, the two engineers advocated "passive" design techniques wherever possible: choosing a larger spacecraft antenna over trying to build a transmitter with higher power, for example. This approach dovetailed neatly with the problem of providing power. If a source other than solar panels were used, the spacecraft wouldn't have to be constantly pointed toward the Sun as it moved in its trajectory toward its destination. So a complex system of maneuvering thrusters wouldn't be needed to maintain the craft's orientation. Instead, it could be stabilized by something as inherently simple as spinning it—exactly like the early Pioneers.

Another major study for a Jupiter mission was put together in 1967 by NASA's Goddard Space Flight Center in Greenbelt, Maryland, which, much like Ames some years earlier, had been looking to become more involved in the solar system exploration business. Goddard proposed an elaborate "Galactic Jupiter Probe" that would not only be the first craft to Jupiter but would also make detailed

studies of the Sun, the interplanetary environment, and galactic phenomena, hence the name. It was a wildly ambitious proposal, perhaps
too much to for a NASA center that had mainly handled near-Earth
missions and satellites up to that time, but a pamphlet detailing the
project spelled out most of the compelling arguments for the mission
in general. "The nation that first flies to Jupiter and beyond," said the
Goddard publication:

"Will be the *first* to investigate the Asteroid Belt between Mars
and Jupiter and the unusual magnetic, radio, and radiation environment of the planet Jupiter.

"Will be the *first* to explore deep interplanetary space beyond the
orbit of Mars.

"Will be the *first* to search for and hopefully locate the region of
interaction of the solar and galactic mediums.

"Will hold the long-distance record for spacecraft communication and flight (about 1 billion miles).

"May well develop information of outstanding scientific importance to our understanding of the solar system, galaxy, and universe.

"Will be developing and maintaining the technological 'edge' required to support planned future interplanetary missions."

And last but not least:

"Will benefit from the prestige which accrues to those who cross
new frontiers *first* and demonstrate their willingness and ability to
undertake difficult tasks important to mankind."

Worthy and lofty goals, to be sure, and definitely shared by every
planetary scientist, including the leading space scientist in the world,
James Van Allen. Van Allen had been fighting for more deep-space
missions for years and had the influence to make his voice heard in
the halls of NASA and Washington. "I was an advocate of missions to
the outer planets, especially Jupiter because of its known enormous

microwave radiation and other types of radio noise which indicated the existence of radiation belts of energetic electrons around the planet," Van Allen explained. Among the many honors, duties, and appointments that supplemented his main stint as a professor at the University of Iowa, he was a founding member of both the prestigious Space Science Board (SSB) of the National Academy of Sciences and the Lunar and Planetary Missions Board (LPMB), the body charged with advising NASA on its planetary programs. "I made such a nuisance of myself in both the SSB and the LPMB that I was appointed to chair a subcommittee of the SSB and also the LPMB panel on the outer planets. . . . The first fruit of our advocacy was the tentative approval of two asteroid/Jupiter missions, subsequently called *Pioneer 10* and *Pioneer 11.*" By 1968, Van Allen and his colleagues on the panels had considered all the studies, questions, and accompanying problems and had come to one conclusion: a mission to Jupiter, and perhaps beyond, was a definite necessity—and now was the time to do it. "Two exploratory probes in the Pioneer class [should] be launched in 1972 or 1973," asserted the board's report to NASA.

Although it had been scrambling for some years to keep some sort of planetary program alive in the midst of Apollo, the Vietnam War, and presidential and congressional indifference, NASA agreed. A relatively inexpensive Jupiter mission would be the perfect way to demonstrate the feasibility of the Grand Tour idea. It would prove that all of the difficulties of such an ambitious mission could be solved; it would blaze the trail to the outer solar system for the next decade at least. There were two main questions: What shape should such a mission take? Who would run it?

With Ames still busy with *Pioneers 6* through *9,* NASA asked JPL and Goddard for ideas on going to Jupiter. JPL, its plate already full, ended up spending too much money for headquarters' taste and soon found itself out of the running. Goddard responded with the Galactic Jupiter Probe but later balked when headquarters insisted on a firm commitment to actively begin project planning, citing a lack of available resources. (However, an internal Ames history tells a somewhat different story. According to the NASA "grapevine," Goddard had been slow in meeting its commitment to support the manned spacecraft communications network, supposedly because of a lack of available personnel; NASA headquarters responded by telling Goddard to "forget the Jupiter mission and use those people to do your job on the communications net.")

If neither JPL nor Goddard could handle the job, Ames was the logical next choice. Headquarters personnel contacted Harvey Allen, now the Ames director, in September 1967 to ask whether the Pioneer office might be interested in taking over the Jupiter mission after the solar Pioneer missions were finished. An intensive back-and-forth discussion ensued on matters of budget, scientific objectives, spacecraft design, and a projected time frame. In a showdown meeting at NASA headquarters in December 1967, Goddard lost its final bid for the Jupiter mission. Aside from the fact that the Ames version would cost less money, Charlie Hall and his Pioneer team had already proven they could do the job. They also had extensive experience with the spin-stabilized spacecraft concept, which was planned for the Jupiter mission for its simplicity and lack of expense.

Finally, some of the more savvy political types at NASA realized another seemingly minor but nevertheless important fact: it was always easier to sell something to Congress as a modification or extension of an already existing program than as something wholly new. Calling the Jupiter mission "Pioneer" in effect gave it the brand-name recognition of a project that was already in operation and running very well, thereby avoiding the nervous doubts and trepidation that tended to affect Bureau of the Budget officials when asked to fund any new endeavor.

More discussions followed throughout 1968, as the spacecraft parameters, mission objectives, and various other details were thrashed out between Ames, NASA headquarters, and other agencies, so that a budget could be formulated. Finally, on February 8, 1969, the Pioneer Jupiter mission was officially approved. Following NASA policy at the time, two identical spacecraft would be launched, one serving as a backup in case the other failed or as a supplement if the first succeeded.

Charlie Hall and his team were going to challenge the Great Galactic Ghoul. If their spacecraft eluded the Ghoul's grasp, it just might make history. Or it might discover that other even more vicious demons awaited beyond. The gateway to the outer solar system was within reach. Whether it was open or forever closed to humanity remained to be seen.

4

The Sole Selection

This time, with the initiation of the Pioneer Jupiter project, Hall had a head start. He already had an experienced team in place, and there would be no need to commandeer any more cafeterias from the Ames populace.

Being able to hit the ground running was a good thing, because there wasn't much time to get the job done. The optimal launch window for a Jupiter mission occurred about once every 13 months. Allowing a reasonable period for the design and construction of the spacecraft and the selection of scientific experiments, the next optimal launch window to Jupiter would be in February to March 1972, a mere 3 years away. After *Pioneer F* (to become *Pioneer 10* after launch, as per custom) was safely en route to Jupiter, its sister *Pioneer G* would follow in the next launch window in April or May 1973. In those 3 years, the *Pioneer F/G* spacecraft (continuing the letter designation series from the earlier Pioneers) had to be designed and constructed; the onboard scientific experiments had to be selected, built by the principal investigators, checked, tested, and delivered; the launch vehicle for the spacecraft had to be readied; communications and data processing requirements had to be determined and prepared; and each spacecraft had to be fully tested, prepared, and finally launched into space. The schedule was almost impossibly tight, leaving no time for wasted effort or waffling over the thousands of decisions that had to be made at each step of the process from project approval to launch. In other words, it was just the way Charlie Hall liked it.

After almost a decade of managing the Pioneer program, Hall had become quite adept at organizing things the way he liked them. Faced with the challenge of sending a spacecraft where none had gone before, Hall knew precisely whom he wanted to build it: TRW, which had done such a stellar job with *Pioneers 6* through *9*. It was a logical choice for several reasons, not the least of which was that the Pioneer team already had an efficient and smooth-running relationship established with TRW. Everyone knew each other; there would be no clumsy shakedown period as the Pioneer team members became acquainted with the contractor's engineers and technicians and each group learned the other's working methods, preferences, and eccentricities. Sticking with TRW for the Jupiter missions also appealed to Hall's fondness for simplicity. Instead of a new contractor designing all the systems from scratch, TRW would be able to adapt and incorporate the existing systems from the earlier Pioneers into the new craft, systems that were already space proven and of undeniable reliability. All of this meant that the process of building and testing the *Pioneer F/G* probes would be faster and less expensive, two more of Charlie Hall's favorite concepts.

The long-established and almost sacred tenets of government procurement procedures, however, dictated that a project as important as the Pioneer Jupiter mission had to be open for contract bids by all qualified parties. TRW's competitors would have to be allowed the opportunity to submit their own ideas and detailed proposals for *Pioneer F/G*, complete with detailed budgets, so that NASA officials could compare and contrast, debate and argue over alternatives, pare down the list of potential contractors, listen to last-minute entreaties by the final candidates, and at long last name the winner and sign the contract. The process was tolerably fair and reasonably democratic, although naturally not immune from politicking and favoritism. The problem, as Hall well knew, was that it took far too long. Taking a year or even two to put out a formal Request for Proposals, collect them, review them, winnow them down, and make a final selection might not matter much when procuring a manufacturer to make widgets for an obscure government project, but wasting so much time on an interplanetary mission wasn't an option. The planets were not going to wait. Miss a launch window, and there would be nothing to do but wait for the next one, which might not happen again for months or even years.

So Hall decided to buck the system. "On the solar probes, we had competitiveness until it came out of your eyeballs, and this was supposedly the way NASA wanted it," he recalled. Although such a competition might have made sense for *Pioneers* 6 through 9, since everyone would be starting from square one in constructing those spacecraft, Hall didn't see the logic in going through the process again in this case, with an experienced and excellent contractor already onboard. "I wrote a long treatise saying that it should be a sole-source selection. In those days that was not a very nice word." "Sole-source" meant just what it said: one contractor would be selected at the outset, without any proposals being accepted and considered from other candidates. It was a gutsy move because it almost inevitably assured a political upheaval. Al Eggers, who by this time was working at NASA headquarters, explained: "If you want to go sole source on something like this, you'd better have a real damn good reason or you're gonna have a tempest arise around you with other competitors saying, 'Hey, we didn't get a chance to compete on that.' It took a lot of courage on Charlie's part to do that and take that strong position. You gotta be a bit of a risk taker to do that."

Hall found an ally at headquarters in Don Hearth, director of NASA's Lunar and Planetary Programs division under which Pioneer would be managed. Hearth's deputy for planetary programs, Robert Kraemer, explained in his book *Beyond the Moon* that Hearth hardly needed to be convinced of Hall's rationale. "Don, a true master of working within government bureaucracy, knew that at Ames he had a strong project manager in Charlie Hall, who worked well with the talented engineers at TRW. Somehow Don convinced his bosses at NASA and all the by-the-book NASA procurement people that *Pioneers F* and *G* were a straightforward evolution of the *Pioneer A* to *E* series and that the two new spacecraft should just be negotiated into Ames's ongoing Pioneer contract with TRW." Kraemer is quite adamant that Hall's and Hearth's insistence on TRW's sole-source selection not only saved years of project time but saved the taxpayers money as well.

Hall succeeded in making his case and got his way. TRW was named as the builder of *Pioneers F* and *G* in the fall of 1969. The company was awarded a cost-plus-incentives contract, which meant that it would enjoy bonuses if it could complete the spacecraft ahead of schedule, keep its weight below the specified parameters, keep down its power consumption, and in general meet or exceed the design specifications. Bernard J. O'Brien, named Pioneer project manager for TRW, was confident despite the great responsibility involved in his company being chosen as a sole source contractor. "We had 100 percent success on the earlier Pioneers," he said. Time was another reason for TRW's selection, in O'Brien's opinion. "The other factor was schedule. For Charlie Hall to get competitive bids and evaluate them and have a source selection would have precluded a 1972 launch. I don't know if anyone else was geared up to tackle the project in that time frame. We had the advantage of working with NASA Ames on the other Pioneers."

Still, O'Brien knew that his work was cut out for him. Even with TRW's considerable advantages, "we didn't have an awful lot of time," he conceded. "There was a million-dollar penalty if we were not ready to launch in February or March of 1972. And I'd have lost my job!"

Herbert Lassen, the tall, bearded engineer who had spelled out the basics of a Jupiter mission back in 1965, had since been placed in charge of designing space missions for TRW and now had to turn his hypothetical concepts into a real, functioning spacecraft. "Herb was an interesting character," O'Brien remembered. "You could never believe he was as smart as he was. He had a way about him that could convince people that he was right. And he could explain things in a logical manner so that even the dumbest of us could understand what he was getting at."

With fellow TRW engineering genius William Dixon, Lassen began to refine and expand on his earlier ideas to develop the final form of the *Pioneer F/G* spacecraft. An intense back-and-forth between Charlie Hall and Lassen ensued. It wasn't an easy process. "He started out laying out various designs and presenting them, and he knew I wasn't liking 'em," Hall recalled. "So then I'd present something I'd been thinking about, and I knew he didn't like *them*. I'd say, 'Keep trying, it'll come.' Finally one day he calls up and says, 'Get down here fast! I got it!' So I go down there and he presented the *Pioneer 10* and *11* that were actually built. It was obvious that it was just a breakthrough."

Lassen had conceived an elegant and ingenious design. "He had thrown all constraints out the window," Hall said. "And some of the constraints were artificial because people assumed something that was not necessarily true." One assumption concerned the fundamental nature of the spin stabilization concept. "Most people in designing a spacecraft have the spin axis go through the center of the antenna so it won't wobble," Hall explained. A wobbling antenna dish could lead to a slight fluctuation in signal power back at Earth. "Herb threw that out and said it doesn't matter. We're talking about 8 centimeters out of about 60 billion kilometers."

Yet in perfect keeping with Hall's keep-it-simple credo, Lassen's innovations were firmly based on proven and tested designs. As Bernard O'Brien pointed out, the *Pioneer F/G* craft "was basically a refinement or evolution of the *Pioneer 6* through *9* design. We used a lot of the same design concepts which, in turn, helped the cost and schedule. I think Charlie was smart in doing that." O'Brien added with a smile, "And of course, I proved it!"

In direct contrast to Lassen's comments in his 1965 paper that "solar power will not be suitable at this distance from the Sun," his initial formal proposals to NASA, including those presented even before the mission was officially approved, were based on the use of solar panels. Since Lassen was certainly a savvy enough engineer to realize the inherent difficulties of this approach, this was probably more of a political consideration than an engineering one. He knew the mission would be easier to sell if the design didn't seem too radically new and unproven. Everyone knew about and understood the idea of solar cells, so proposing their use avoided the doubts and arguments that a new and untested power source would evoke.

Now that the mission was a definite go, however, Lassen returned to his original ideas. The *Pioneer F/G* craft would be powered by radioisotope thermoelectric generators, RTGs for short, devices that generate electricity from thermocouples using the heat produced by small capsules of decaying plutonium-238. RTGs had been used for some earlier satellites but were considered unreliable for long-term use, especially for a deep space mission such as Pioneer. "At first they looked like they were going to be one big pain in the ass," Hall admitted frankly. "They weren't developed far enough long yet."

The Atomic Energy Commission and the Teledyne Corporation had been steadily working on improving RTG technology in the meantime. A new unit called the SNAP-19 (for Systems for Nuclear Auxiliary Power) designed for a much longer lifetime was ready, and the AEC was looking for a way to put it to use. One evening at the end of the workday at Ames, a visitor from the AEC dropped by Hall's office. "We must have spent about 2 or 3 hours in my office talking," Hall said. They were talking about the SNAP-19 and the possibility of using it to power Pioneer to Jupiter. "He was really anxious to get a spacecraft to put a unit on, and ours was the only one at that time of an interplanetary nature. He said we'll build the prototypes free, and all you have to do is pay for the flight units. Well, this was a big bonus—saving us 15 million bucks. I couldn't turn that down!" Four

SNAP-19s would provide about 155 watts of power, more than enough for the nominal mission to Jupiter.

According to Robert Kraemer, ultimately NASA didn't even have to pay for the RTG flight units. Upon discovering that the AEC's director of radioisotope programs lived across the street from him, Kraemer managed to convince him that "it was in the AEC's best interest to not only develop RTGs for spacecraft but also to provide the flight units at no cost to NASA." The arrangement turned out to be so successful that it was continued for the later Voyager and Viking RTG-powered missions.

The switch from solar cells as the ostensible power source to RTGs might have made perfect engineering sense but wasn't without serious problems. Even if the RTGs themselves were free, their use required some major design changes in the spacecraft itself. Stray radiation from the RTGs couldn't be allowed to interfere with the scientific instruments or the spacecraft's operating systems, which meant that extra shielding and other modifications were essential. Design changes ate up two resources that Charlie Hall could ill afford to squander: money and weight. But Hall had anticipated such difficulties. "Fortunately, I had about 50 or 60 pounds of contingency that I could dole out when people got into problems like this. For problems like this, you had three choices: don't do anything, spend a lot of money, or add a little weight in the right places."

Lassen's final Pioneer design was compact, simple, and utterly basic, yet perfectly suited to the mission requirements. A 9-foot parabolic high-gain antenna was mounted on a 14-inch-deep hexagonal instrument compartment containing the operative "guts" of the spacecraft. Another smaller hexagonal compartment on one side of the main compartment housed most of the 11 scientific instruments. Three small pairs of thrusters were mounted along the edge of the main antenna for attitude and spin control. And three long booms, equally spaced at 120 degrees around the spacecraft circumference, extended from the main compartment. Two of these booms, each about 10 feet long, held the four RTGs, two on each boom, at a com-

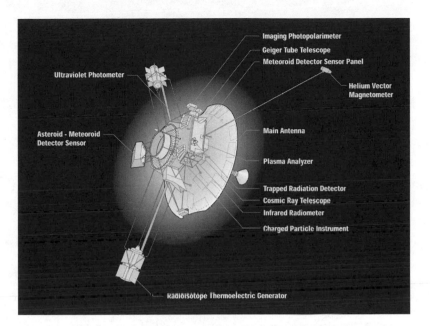

Ultraviolet Photometer

Imaging Photopolarimeter

Geiger Tube Telescope

Meteoroid Detector Sensor Panel

Helium Vector
Magnetometer

Asteroid - Meteoroid
Detector Sensor

Main Antenna

Plasma Analyzer

Trapped Radiation Detector

Cosmic Ray Telescope

Infrared Radiometer

Charged Particle Instrument

Radioisotope Thermoelectric Generator

The final *Pioneer 10* and *11* spacecraft, with its complement of science instruments.

fortable distance from the rest of the spacecraft. The third, about 17 feet long, carried one of the scientific experiments, a magnetometer, keeping it well away from any possible magnetic interference from the spacecraft or its other instruments. The entire craft weighed only 570 pounds and, minus the extended booms, was about 9 1/2 feet long by 9 feet wide, small enough to fit into an average living room. And more importantly, it was small and compact enough to fit into the nose cone of the Atlas-Centaur booster that would carry it forever away from home.

Keeping Pioneer so simple, small, and lightweight entailed some significant sacrifices. Making the craft spin stabilized like the earlier Pioneers saved a great deal of weight and complexity, of course, but it also meant that Pioneer's instruments would be constantly spinning. For the scientific experiments that measured particles and fields, this was actually an advantage because it allowed them to scan large areas quickly and build up a data map of whatever patterns might be present. For picture taking, however, it was a definite problem. Taking pictures wasn't a prime mission objective, and many scientists had little or no interest in taking in the Jovian sights. But to journey all the way to Jupiter for the very first time and not be able to *see* anything would be ludicrous, and not very popular with the American taxpayers, who didn't care about magnetospheres and charged particle radiation but were ultimately footing the bill.

Fortunately, one of the principal scientific investigators who would soon be chosen by NASA, Tom Gehrels of the University of Arizona, had an answer. Pioneer couldn't carry a conventional camera, not only because of the weight but because the craft's spinning would make it useless. But Gehrels's instrument, called an imaging photo polarimeter or IPP, would be able to create pictures of Jupiter for the press and public with the aid of some special computer processing of the data back on Earth. Using a technique called spin scan imaging, the IPP could scan strips of its field of view in blue and red light, in a successive pattern that could be assembled by computer into full images. The IPP images wouldn't be of the high quality and resolution possible with an actual television camera, but they would be obtained from a distance closer than anyone had ever seen Jupiter. That was impressive enough for this first mission.

Another factor in Pioneer's simplicity had an even greater bearing on the nature of the mission and how it would be run. There was no room or power for an onboard computer in which to store complex instructions and data and to control the spacecraft when it was out of touch with Earth. As Bernard O'Brien said, "The spacecraft had no 'brain.' All the commands had to be sent from Earth. There were some simple reliability solutions which triggered automatic changeover if one of the redundant units failed. That's about as smart as it was."

"It was a real-time spacecraft," said Dave Lozier, who worked on trajectory computations and launch vehicle preparations for the Pioneer team. "Everything you wanted to do was commanded from

Earth. It had a five-command storage feature so if we had a power problem and had to shut instruments off in an emergency, it would just cycle through these five commands. Anything we wanted to do —fire the jets, turn an instrument on and off—we had to load a register and execute it." Lozier explained that the decision to exclude an onboard computer was both practical and consistent with the Pioneer credo of simplicity and reliability. "Back then they were just starting with semiconductor devices. It was part of the philosophy to keep the spacecraft simple and uncomplicated by a lot of electronic wizardry. Also there was a mass accommodation. It takes a lot of energy to get to Jupiter. You don't want to load the spacecraft up with too many boxes."

Because the spacecraft lacked a "brain" and thus wasn't self-sufficient, the issue of communications would be absolutely vital. With the light-speed-dictated time lag between message transmission and reception ever increasing as Pioneer sped away from Earth, spacecraft maneuvers and possible problems had to be anticipated well in advance. And over the immense distances to the outer solar system, Pioneer's 8-watt signals would be a mere whisper, lacking enough energy to power the dimmest night-light and pushing the extremely sensitive antenna dishes of NASA's worldwide Deep Space Network to their limits.

Like almost any other complex engineering problem, designing a spacecraft is an endless series of tradeoffs and compromises. Fortunately for the Pioneer mission, circumstances had conspired to bring together a group of individuals—Charlie Hall, Bernard O'Brien, Herb Lassen, and many more—who were virtuosos of compromise.

"A FIERCE, ALMOST BLOODY COMPETITION"

As Herb Lassen, Bill Dixon, and the rest of the Pioneer TRW contingent began designing and building the Pioneer spacecraft at TRW headquarters in Redondo Beach, California, some of the most prestigious and the most unconventional scientists from all over the United

States were devising, designing, and putting together the scientific experiments that provided the purpose of the mission's existence. For some of the scientists it was another in a long career of space missions; for others it was the very first time they had been given the opportunity to place an experiment on a NASA space mission. For all of them it would be a long and arduous journey from idea to reality.

The NASA Office of Space Sciences had issued a formal Announcement of Opportunity (AO) to the scientific community on June 10, 1968, after Ames had been chosen to administer the project 8 months before its formal approval by NASA. "The space flight missions outlined herein have not been authorized," warned the AO in its very first sentence. "They are currently under consideration by Congress and the Executive Department. . . . However, in order to insure maximum scientific participation and to proceed in an orderly way with the selection of scientific experiments, it is necessary to issue this document at this time in anticipation of such authorization." Scientists seriously interested in participating in the mission, which at this point was defined only as "adapt[ing] Pioneer-class spacecraft to go out beyond the orbit of Mars, through the asteroid belt, and toward the orbit of Jupiter," were invited to submit letters of inquiry detailing possible experiments; a later meeting of interested parties would firm up the mission objectives and spacecraft parameters, after which NASA would invite formal proposals.

Details were sketchy at this point. The AO noted that TRW was conducting a study "to investigate the problems associated with adapting presently designed Pioneers for Asteroid/Jupiter missions while maintaining spacecraft modifications to a minimum," words that no doubt warmed the cockles of Charlie Hall's heart. The spacecraft "will incorporate as many of the basic subsystems from *Pioneer VI* as possible," said the document, although it was "anticipated that its external appearance will differ considerably from *Pioneer VI*" due to the different mission objectives. Almost as an afterthought, solar cells were specified as the power source, a detail that didn't escape the attention of some of the shrewder scientists and soon played a significant part in the final selection of the experiments.

James Van Allen noted that the AO "borrowed almost all the scientific rationale for the mission from the earlier report of my panel." The AO cautiously pointed out that "close fly-bys of Jupiter are also being considered, though the technical feasibility of such fly-bys during the early missions has not yet been established." Still, the document invited suggestions for spacecraft trajectory and the "desired distance between the spacecraft and Jupiter at closest approach."

The response to the AO was swift and overwhelming. More than 150 Letters of Inquiry had poured into NASA headquarters (along with copies of every one of them to Charlie Hall) by August 26, 1968, when the informational meeting was held at Ames. Apparently it wasn't quite as illuminating as had been promised, at least for some participants. Joe O'Gallagher, who represented John Simpson's University of Chicago team at the gathering, later complained in a memo to Simpson that "very little has been firmly decided by the project" with regard to the specifics of the mission or spacecraft design, or even whether it would be powered by solar cells or RTGs. O'Gallagher was surprised at the size of the competition, however. "Probably as many as one-third to one-half of all the experiments proposed are unrealistic or inappropriate for this mission," he observed, "but it is clear that competition to be on board will be strong." By October, NASA was ready for formal detailed proposals. The deadline given was December 2, 1968.

Thus ensued what Simpson, tongue only partly in cheek, characterized as "a fierce, almost bloody competition." Getting an experiment on the very first spacecraft to penetrate the outer solar system would be an enormous, unprecedented coup for any space scientist. It meant that he (in those years space scientists were invariably male) would be the first to make observations and collect data from a hitherto unexplored frontier. He would establish the foundation upon which all future work would be based. Just as the Pioneer space probe

would pathfind for all the spacecraft that followed, so the Pioneer
principal investigator (PI) would blaze the trail for all the scientists
and graduate students who would work on future space missions.
Disregarding the prestige that such an accomplishment would bring
to oneself and one's academic institution, and the fact that a successful experiment would undoubtedly lead to further opportunities on
later NASA missions, there was the most basic of all motivations for
any scientist: boundless curiosity. For many scientists, Pioneer offered
the first chance to answer questions that had plagued them from the
beginnings of their careers.

They had only months to phrase those questions in the form of a
tangible scientific instrument that would not only meet all the strin-
gent limitations of the spacecraft's design but also survive and work
perfectly in the hostile environment of deep space. To succeed, a space
scientist had to be a skilled engineer as well as a scientist and as handy
with a soldering iron as with a slide rule. As John Naugle, former
NASA associate administrator of space science, noted in his memoir:
"Unlike some of their colleagues, space scientists could not conceive
of experiments on Sunday afternoon as they sat on their patios sip-
ping a scotch and soda, go into their laboratories on Monday, build
the apparatus, take data a month later, and send off a paper to the
Physical Review before the year was out. If they pondered a question
on their patio, they had to figure out how to build an apparatus that
could survive the stress of a rocket launch, operate unattended for
months or years, and accurately transmit the data necessary to answer the original scientific question." It was a form of scientific ex-
perimentation and exploration that allowed no margin for error, no
room for mistakes. If your instrument didn't work after launch, it
was worse than useless: it was dead weight, contributing nothing to
the mission and perhaps even interfering with it.

If all that wasn't enough pressure to keep a potential Pioneer PI
up at night, there was the thought of the competition. Those unfamiliar with science and scientists except as portrayed in school text-
books and television documentaries tend to view the scientific

enterprise as an ivory-tower collegial undertaking among dedicated scholars, something pure and noble, untainted by the pettiness and competitiveness and similar less uplifting human motivations that infest other endeavors such as, say, the business world or the entertainment industry. In reality, science can be just as rife with the less lofty human emotions and behavior as any other activity; the difference is that the inherent checks and balances of the scientific method, strict peer review, and experimental verification tend to smooth out the rougher angles of human nature, at least in the larger scheme of things. In the scientific trenches, however, the rivalries, competition, and resentments can be just as real and as overwhelming as in the corporate boardroom. While open vindictiveness is rare, competition among universities and other institutions, laboratories, research teams, and individual scientists is a fact of life. Scientists are continually forced to compete for funding, faculty positions, project approvals, and, of course, the coveted slots on a space mission.

There were going to be only 13 of those slots for the Pioneer spacecraft. By the December 1968 deadline, NASA had received 75 formal proposals, each of them hundreds of pages long, detailing a scientific rationale, instrument design parameters, budget requirements, and every other aspect of the proposed experiment. Some could be eliminated from consideration almost immediately because they were frankly not in line with the mission objectives; as the NASA request had stated: "Investigations such as space biology or the observation of extra solar system objects which do not contribute directly to meeting the [mission] objectives will not be considered for assignment to *Pioneer F/G*." Other proposals were obviously too expensive or impractical. Many more were apparently worthy but so similar in their essential characteristics that it was hard to distinguish among them. A NASA review committee plowed through the proposals, weighing them, comparing and contrasting, laboriously winnowing the avalanche down to a more manageable 25. Out of this penultimate cut would come the 13 experiments selected to go to Jupiter and beyond.

The final selection process was unavoidably merciless. Simpson recalled: "They decided they would have a 'shootout' hearing on these in which all the investigators would be present to make a 20- or 30-minute presentation of your proposal, and the others would attack you." Those who survived this grueling ordeal, in effect forced to prove their proposal's superiority to their competitors, would be the victors. It happened in Washington in early 1969. To make it even worse, the decisions weren't even announced to the scientists at that time but only anxious weeks later. "I awaited NASA's formal decisions with a mixture of optimism and pessimism," James Van Allen recalled. "I had served on many peer review committees and was acutely aware of the nature of their decision-making process. Often, a mere shrug or a few words by an influential member would make or break the professional future of a research team." The fact that such trepidations were felt even by a scientist of the stature of Van Allen, a man who had literally founded the modern discipline of space science and had flown more instruments into space than anyone else, is somber testimony to the ruthless reality of the selection process.

Before the spring of 1969 dissolved into summer, Van Allen, Simpson, and 11 other research teams learned they had made it. They were going to Jupiter, at least by proxy through their experiments. There would be 11 scientific instruments onboard the craft; two of the experiments actually used the spacecraft itself and didn't require specialized equipment. For some of the newly named Pioneer PIs, experience and a bit of foresightedness had been the deciding factors. Simpson, for example, had anticipated the switch from solar cells to RTGs, even though the official decision to do so wasn't made until the fall of 1969. He had cleverly designed his charged particle experiment to allow for stray radiation from a nuclear power source, testing early versions of his instrument by exposing them to the same types of radiation that RTGs might emit.

"I felt that only RTGs would do the job, so I started the design for an unknown mission which turned out to be very close to *Pioneer 10* in 1963," he explained. "I arranged that an instrument I designed for an *IMP* [*Interplanetary Monitoring Platform*] satellite be taken to Goddard Spaceflight Center and at my request the Office of the Navy would deliver a transit satellite RTG. They delivered it to Goddard and I went down with my engineer and we made a study of the background effects on our instrument, which of course was not precisely the instrument we would eventually fly [on *Pioneer 10*], but it provided us with information on particle background and how we could eliminate the background by raising various thresholds for detection and so forth to a level where we wouldn't have any problem. So 4 $^1/_2$ or 5 years later [with *Pioneer 10*] when it came time for the competition, my competitors were waving their hands about how they would solve the problem and I was able to show that we had solved the problem and our detector would do the job. That's how we won the competition."

Van Allen had done similarly, his vast experience in designing detectors of cosmic rays and other radiation making him an expert in finding ways to avoid unwanted detections. For other PIs a bit of luck had entered the equation. The University of Arizona's Tom Gehrels recalled that his imaging photo polarimeter had been narrowly chosen over its closest competition mostly because his instrument was pointable and the competition's wasn't.

With the onboard experiments selected, contracts, budgets, and schedules were hammered out between the Pioneer project office and each of the PI teams by Hall and Joseph Lepetich, his liaison to the scientists. Lepetich found that he sometimes had to do a fair amount of whipcracking in order to keep some of the scientific teams on budget and on schedule. It was always something of a delicate balancing act because while the focus of each PI was naturally on his own small piece of the project—his experiment—all of the scientific instruments had to be seamlessly integrated to work together in a single spacecraft. An alteration in one instrument's size, power requirements, weight, or any other parameters meant that something else had to be changed to accommodate it.

Throughout 1970 and into 1971, the Pioneer PIs worked relentlessly. Each team first had to construct a prototype instrument to prove the workability of their design, then a design verification unit for use in testing, and finally two actual flight instruments for the *Pioneer F* and *G* spacecraft. Frequent meetings with the Pioneer team at Ames Research Center ensured that everyone remained on the same page. "There were massive negotiations between the project and the family of investigators on things like power and weight," Van Allen said. "Charlie had a very conservative view on all of those in the most constructive sense of what was feasible and could be reliably done within the capabilities of the budgets." But although Hall presided with a firm hand, he managed to keep everyone happy. "He was our best friend," remembered Van Allen. "There was never any doubt that he was a representative of the federal government, but at the same time he was one of the most intelligent and responsive project managers I've ever worked with in the NASA system. We had quite diverse scientific objectives, but Charlie was a prime advocate of keeping it clean from an engineering standpoint."

Conflicting objectives, of course, meant that more often than not, a comfortable middle ground had to be found. "There were lots of compromises made," Van Allen agreed. "We would have loved to have more telemetry capacity and more power available and to add more weight. All those things were hashed out in a series of group meetings in which we traded back and forth, one thing for another. Very constructive, but also rigorous negotiating sessions."

Tom Gehrels recalled one example of Hall's leadership style that solved a particularly thorny problem during a scientific planning conference. With 11 scientific instrument packages aboard Pioneer, and a slow data rate of 1,024 bits per second, there was no way that all of the PIs would be able to have complete coverage during the crucial encounter phase. The particles and fields specialists, particularly Van Allen and Simpson, wanted priority. Understandably so, since their instruments would be the busiest and provide the most information during the Jupiter flyby. That meant that other experimenters, including Gehrels, would be left more or less out in the cold. But how could a scientist like Gehrels, on his first space mission, challenge senior colleagues like Van Allen and Simpson, who not only had much more spaceflight experience but were largely responsible for the mission's existence in the first place?

"Tension was building in the packed conference hall," Gehrels wrote in his memoirs. "How would one resolve this? Most other programs would have set up a committee, with panel meetings and reviews by competent referees, to do a study—25 copies or more—at useless expense and loss of time. Not Charlie Hall, who thrived on problems. First he declared a coffee break to relieve the tension, suggesting I chat with Van Allen and Simpson. After a while he quietly joined us and said, 'How about each of you—imaging and particles and fields—getting half of the data link?' What was there left to say? If he had offered 60 percent, we would have fought for 65 percent. But half each had a touch of fairness, equal rights, and King Solomon's wisdom. So we proceeded, relieved, to the next problem."

The PIs also worked closely with the engineers who were actually building the spacecraft. "We had a lot of sessions with the TRW engineers, who were an extremely competent group," said Van Allen. "A lot of give and take with the engineering staff at TRW under Charlie's general oversight. We got along with them great. TRW was very receptive to everything we wanted to do. They were under fundamental restraints with the permitted dimensions of the spacecraft, the limited weight, all kinds of thermal considerations, power available from the RTGs, and so forth. But the constraints were very intelligently applied."

Inevitably, each PI had to stop refining, redesigning, and trying to improve his instrument and the design had to be "frozen" so that the spacecraft could be finished and prepared for launch. Jack Dyer, Hall's mission analysis chief, recalled that the point of "freezing" an experiment design was always hard on the scientists but that Charlie Hall, while sympathetic, would brook no arguments when that time came. "He saw other projects that just ballooned in cost," Dyer said. "He recognized that the scientists would improve their instruments, but he knew he had to restrain that, otherwise he'd get into the same situation of exploding budgets. Also, those kind of things affect schedule as well as cost and Hall needed to meet the planetary schedule, which won't wait for any of us."

From a scientist's point of view, though, letting go of an experiment is never easy. The temptation to endlessly improve and refine is overwhelming, and not just because a scientist might have an affinity for tinkering. As Van Allen said, "Deficiencies of design and construction of space equipment are, to a considerable extent, incurable in flight. . . . Moreover, we usually work on tight time schedules and our equipment must meet mechanical, electrical, and thermal specifications far beyond those needed for laboratory equipment, and all within firm constraints on mass, power, shape, size. We called these constraints the battle of grams, milliamperes, and cubic centimeters." A space scientist's job was also hardly sedentary: "Numerous transcontinental flights for purposes of testing and integration are a way of life." And even after the final tests on the ground are completed and the PI's instrument is incorporated into the spacecraft and finally launched, it's never really over. "The first phase of the 'final examination' is the in-flight checkout of one's equipment soon after launch; but the exam continues every day throughout the mission."

A May 1970 conference at Ames for the press and the scientific community spelled out for the world at large the ambitions of Hall and the PIs. In less than 2 years, *Pioneer F* would leave for Jupiter and parts beyond. About a year later its sister craft, *Pioneer G/11*, would follow.

It was almost no time at all, especially for those who had to make it work.

5

Countdown and Controversy

y the beginning of 1971 the Pioneer Jupiter project was moving along nicely. TRW was building the spacecraft at its headquarters in Redondo Beach, the principal investigators (PIs) were working on their science instruments, and Earth and Jupiter were drawing inexorably closer to the ideal alignment for a 1972 launch.

There were still some troublesome issues to be resolved, however. The radioisotope thermoelectric generators (RTGs), for one, were still a relatively new and not fully proven technology, something that became obvious as testing of the spacecraft and its instruments began. The switch from solar cells to RTGs caused no end of headaches for some of the PIs, some of whom found themselves in the position of having to redesign their instruments almost from scratch. Even those who had anticipated the switch, such as Simpson, discovered as they tested their instruments that the RTGs would cause unacceptable interference with their experiments without further modifications such as extra shielding and circuit redesign. The Atomic Energy Commission and Teledyne strived to reduce impurities in the radioactive ma-

Pioneer 10 undergoes final checkouts at TRW in December 1971 before being shipped to Cape Kennedy.

terial within the RTGs, which would reduce stray radiation. Another problem arose when Teledyne discovered that moisture trapped inside the RTG radioactive fuel capsule was seriously affecting the nuclear reaction and thus sharply reducing the practical lifetime of the RTGs, not to mention making the fuel brittle and subject to disintegration at the slightest shock. Changing the insulation material inside the RTGs solved the problem but not before causing some fears about missed launch windows.

Other problems were more mundane, if not less critical. Simpson, for example, found that his charged particle experiment seemed to attract the unwanted attentions of clumsy technicians to an unusual degree. Within a 6-month period, windows on his particle telescopes were punctured five times, either when the instrument was undergoing testing at Ames or during spacecraft construction at TRW. Each time Simpson's instrument had to be repaired and retested to ensure that it was still functional for the mission. Understandably miffed, Simpson pointed out in a letter to Joe Lepetich that although the windows on his instrument were tough, they were "not immune to direct puncture by human intervention." He also noted that on a previous space mission (the *Orbiting Geophysical Observatory 5* or *OGO-5*) "we had a window of similar thickness of the same material about four inches in diameter and had no losses throughout the entire program."

While RTGs were being refined and instrument windows repeatedly punctured, trajectory specialists such as Jack Dyer were considering the best course to Jupiter. Pioneer would not, after all, be hanging around the Jovian neighborhood. One quick pass by the planet at a speed of thousands of miles an hour was going to be the all-too-brief culmination of the journey. In that brief period of closest approach, the Pioneer spacecraft had to accomplish its prime mission objectives—and survive long enough to transmit its findings back home. How close could the craft pass and still survive Jupiter's intense radiation? What angle of approach would provide the best sampling of the radiation belts and magnetic field distribution? What trajectory would yield the best view of the Jovian cloud tops in a perspective impossible to see from Earth? Would it be possible to take a bonus look at one or two of Jupiter's moons?

The more information that could be captured by the first Pioneer flyby, the better the second Pioneer Jupiter mission, as well as future missions by other craft, could be effectively planned. But all would be for naught if the spacecraft was so badly fried by Jovian radiation that it suffered "radiation death." And even if the spacecraft survived, the radiation might affect its systems, generating false commands that couldn't be corrected in time because of the 90-minute communications delay between Jupiter and Earth. The trajectory questions fueled some lively debates, particularly among the PIs whose experiments and the data to be gained from them depended on the final answers, as launch time drew nearer. Should *Pioneer 10* fail, only the potential backup by its sister *Pioneer 11*, which would get one more shot at Jupiter, provided some reassurance. The hard fact was that until the first spacecraft passed Jupiter there was really no way to be certain about any of the dangers; only a best guess was possible.

One aspect of the trajectory *was* certain, however. From whichever direction Pioneer encountered Jupiter, it would do so at such a high velocity that, coupled with the gravitational kick Jupiter would impart to it, the craft would be sent hurtling out of the solar system, never to return. Everything else launched before by human beings, even the early Pioneer probes that were lost forever, would never depart the gravitational realm of the Sun and would remain trapped within the solar system. *Pioneer 10* would be the first humanmade object to someday enter interstellar space.

On a purely practical level, this fact wasn't very significant. No one really expected that Pioneer would live long enough to actually enter interstellar space in any condition to send useful data back to Earth, and even if it did, the craft wouldn't encounter another star system for uncounted millennia. The spacecraft was designed to make it to Jupiter, nothing more; that in itself would be enough of a feat. Some of the Pioneer scientists hoped that the craft might hold out long enough to return some interesting data from the fringes of the outer solar system, but the engineers thought otherwise. Even if the RTGs held out that long, useful communications at such distances would be virtually impossible. Yes, the thought that Pioneer would eventually leave the solar system was interesting and would chalk up another notable first in a mission already loaded with firsts, but it wasn't of any serious value from an engineering or scientific standpoint.

To some with a decidedly more philosophical bent, the prospect of Pioneer leaving the Sun's family behind was of definite value. It led to an eleventh-hour effort that would give Pioneer one more job that Charlie Hall, Herb Lassen, and the rest of its creators hadn't ever envisioned—and would lead to more public controversy than had ever before accompanied the launch of an unmanned space mission.

THE NAKED PEOPLE

Eric Burgess, an expatriate English science writer for the *Christian Science Monitor*, gazed through thick glass at the *Pioneer F* spacecraft undergoing tests in a deep space simulator at its birthplace, TRW Systems, in late 1971. Locked alone in a vacuum chamber for an entire week, the craft was running solely on RTGs under conditions of deep cold and almost complete darkness, just as it would in space. In a few weeks, Burgess knew, the spacecraft would be carefully packaged in a large container (reusable, of course, following standard Charlie Hall management and budgeting philosophy) and shipped to Cape Kennedy, where it would be prepared for launch. After leaving Earth it would never again be seen by human beings, except as lines on computer screens and sheaves of data printouts.

"I thought, my goodness, that spacecraft is going to escape from the solar system," Burgess remembered. "And it would be a shame to send it out without some kind of a message to extraterrestrials." *Pioneer 10* would be humanity's first ambassador to the universe, but no one seemed to be giving that particular fact the proper attention. It seemed to Burgess that there had to be some way not only to acknowledge it but also to use it. Perhaps some kind of simple message could be placed aboard Pioneer that would say something about where it came from and what sort of beings sent it on its journey. "I'd seen that we'd put plaques about the president and Congress and all sorts of people on some of the other vehicles that had gone to the Moon," said Burgess. "I thought, that's ridiculous, these people will be forgotten in a million years. We need a message that will last."

Burgess wasn't sure how to approach NASA with such an idea, though. He talked it over with planetarium expert Richard Hoagland and *Los Angeles Herald-Examiner* reporter Don Bane. "We can't do it. We're newspaper people," Burgess said to them. "But I know Carl Sagan, and in fact I was with Carl Sagan just a couple of nights earlier, and he was telling how he and Frank Drake had tried to get a message to extraterrestrials by radio." Sagan was the perfect individual to champion the concept, not only as a prestigious planetary scientist who worked on NASA's unmanned missions but also as a famous (some would say infamous) proponent of SETI, the Search for Extraterrestrial Intelligence. Upon hearing about what Burgess and his colleagues were thinking, Sagan's "eyes lit up," Burgess recalled. Sagan used his influence to win quick approval for the idea from NASA.

Over coffee at a meeting of the American Astronomical Society in San Juan, Puerto Rico, Sagan enlisted Frank Drake of Cornell University's National Astronomy and Ionosphere Center (and also a SETI enthusiast) for the cause. Sagan envisioned some kind of metal plaque engraved with a message, which would not only be lightweight but also last for millions of years in interstellar space. But just what should the plaque say? Drake suggested "an interstellar postcard" consisting of a diagram of the solar system, a map showing Earth's location in the galaxy, and an element that was to cause more trouble than either man suspected—pictures of human beings.

The plaque had to be designed and made quickly because every component and every ounce of weight added to the spacecraft had to be strictly accounted for and the February-March 1972 launch window had to be met, whether or not any "postcard" it might carry was ready. The last-minute addition also meant that the probe's design had to be "unfrozen" in order to accommodate the extra mass of the plaque, because the spacecraft's maneuvering thrusters would have to

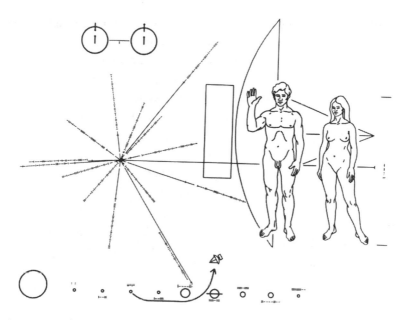

The Pioneer plaque.

be realigned to compensate for the consequent slight change in the craft's center of gravity.

Sagan's artist wife, Linda Salzman Sagan, drew line figures of a man and a woman, the man's hand raised in friendly greeting. Originally, she drew the humans holding hands but then realized that an extraterrestrial being might think the figure was a single creature with two bizarre appendages instead of two separate beings. Striving to depict an idealized and generic human form, Linda tried to make the man and woman as racially neutral as possible. Meanwhile, Sagan and Drake devised a map that plotted the Earth's galactic location with respect to the galactic center and 14 prominent pulsars (stars that generate regular pulses of strong radio waves). In the top left of the plaque, they placed a key to the measurements depicted in binary symbols in the rest of the diagram: a representation of the hyperfine transition of the neutral hydrogen atom, a natural phenomenon that generates a radio wave of 21 centimeters wavelength, a physical "yardstick" presumably known by any technological civilization. With a schematic drawing of the Pioneer spacecraft (to scale with the human figures), and a diagram of the solar system (with a tiny Pioneer arcing away from the third planet, its antenna pointed toward home), the design was complete.

Two identical 6 × 9 inch gold-anodized aluminum plaques were engraved—one for *Pioneer 10* and the other for its sister craft, *Pioneer 11.* The plaque was firmly affixed to two of the struts connecting Pioneer's main antenna dish to the body of the spacecraft, facing inward to minimize erosion of the engraving from micrometeorites and interstellar dust during the eons it would drift through space until its message was read—if it ever was.

Sagan, Drake, and the Pioneer project team knew, of course, that neither they, nor probably any of their remote descendants, would ever witness any response to the plaque's message. Such a response, if

A view underneath *Pioneer 10*'s main antenna dish, showing the plaque mounted facing inward for protection against erosion by micrometeoroids and dust.

it ever came, would be in the far distant future, long after Pioneer, its plaque, and its original mission had been forgotten.

They were wrong. The message of the Pioneer plaque elicited a huge and extremely vocal response. Not from its intended recipients but from an entirely unexpected quarter.

The Pioneer plaque went public in February 1972, shortly before *Pioneer 10* was set to leave Earth forever. Newspapers, magazines, and television news shows played up the fact that the plaque would be the first artifact sent into space solely to communicate with alien beings. "We're fascinated by scientists' attempt to communicate with distant civilizations," proclaimed the *Birmingham News*. The *New York Times* was more philosophical in its comments: "Despite the uncanny mastery of celestial laws that permits man to shoot his artifacts at the stars, we find ourselves still depressingly inept at ordering our own systems here on earth. . . . The marker launched into space is at the same time a gauntlet thrown down to earth: that the gold-plated plaque convey in its time the message that man is still here—not that he had been here." Sagan and Drake made the media rounds, explaining the science behind the message and the odds for the existence of extraterrestrial life. "We do not know if the message will ever be found or decoded," Drake said in *Time* magazine. "But its inclusion on the Pioneer spacecraft seems to us a hopeful symbol of a vigorous civilization on Earth."

But the news media faced one rather embarrassing dilemma with this particular story: the man and the woman on the Pioneer plaque were depicted in the nude. It was perfectly logical, as far as Linda Sagan was concerned. The plaque was intended to be scientifically accurate in all respects and to provide as much helpful information as possible to alien beings, who would be wholly unfamiliar with human biology and anatomy. Since the natural state of the human being is nudity, she drew the figures without any clothing. Not only was it biologically truthful, her decision solved the pesky problem of just what style of clothing should be chosen to represent all of humankind.

Robert Kraemer, then at NASA headquarters, admitted in his memoirs that "when the plaque design was submitted to NASA headquarters for approval I must confess that I was a bit nervous about it. Linda was a skilled artist and her naked human figures were very detailed and realistic, as they needed to be. It seems a bit silly today, but at the time I feared that some taxpayers, the true owners of the spacecraft, might label it pornographic. My boss, John Naugle, had no such fears and approved the design but with the one compromise of erasing the short line indicating the woman's vulva. (The poor extraterrestrials are probably going to be puzzled by the functional differences in anatomy between the two figures.)"

Unfortunately, human nudity was anathema in the American media. It didn't matter that the figures on the Pioneer plaque were hardly engaged in anything more erotic than the act of standing. True, the male figure's hand was occupied, but only in being raised in a friendly gesture. The simple fact of their nakedness was enough for the image to be considered pornographic in many supposedly otherwise enlightened circles.

Still, Pioneer and its plaque were big news. The spacecraft itself was no problem; it wasn't even pretty, much less erotic, so large photos or even intimately detailed diagrams could be published without trepidation. The plaque was another matter. How could a newspaper or TV show do a story on the plaque without *showing* it?

The *Philadelphia Inquirer* solved the problem by summarily erasing the woman's nipples and the male's genitals from its picture of the plaque, citing the need to "uphold community standards." Assistant managing editor Andrew Khinoy later explained to a *Wall Street Journal* reporter (whose own publication had taken the easy way out by not showing the plaque at all): "What they're ready to accept in outer space, they're not ready to accept in Philadelphia, at least not on the front page of their newspapers." The *Chicago Sun-Times*, on the other hand, was more conflicted in its own censorship: it published an unaltered picture in an early edition, then airbrushed out the offending anatomical features, one after another, in each subsequent edition of the same day. Other newspapers and broadcast outlets took the cruder but nonetheless effective step of blacking out the naughty bits with duct tape. The *Wall Street Journal* wryly remarked that "whoever finds the plaque seems certain to be puzzled about the purpose of the male organ and to be left in the dark about human reproduction. Which, apparently, is just where a lot of earthlings would like to have been left themselves."

No doubt emboldened by its neighbors in Hollywood, the universally acknowledged Sodom of the United States, the *Los Angeles Times* showed the Pioneer plaque complete and uncut, for which it suffered the indignant slings and arrows of numerous highly affronted citizens. "I was shocked by the blatant display of both male and female sex organs on the front page of the *Times*," wrote one. "Surely this type of sexual exploitation is below the standards our community has come to expect from the *Times*. . . . Isn't it bad enough that our own space agency officials have found it necessary to spread this filth even beyond our own solar system?"

A somewhat more bemused correspondent wrote to *Natural History*, remarking on the possibility of aliens finding the plaque: "I am sure that within a thousand years they can interpret enough of it to know there is a nudist colony somewhere in outer space."

Editorial cartoonists had a field day. One cartoon showed two little green men smiling at a couple of scientists. One scientist tells his just-arrived colleague: "They say they got our message in the Pioneer spacecraft, and they've come to see the naked women." Another cartoon portrayed two well-dressed and very human-looking aliens regarding the plaque, with one observing that "the Earth people are evidently very similar to us . . . except that they don't wear any clothes!"

Others found the nudity of the plaque's figures their least offensive aspect. Some feminists complained abut the female figure's "passivity": she was smaller than the male, with both arms hanging limply at her sides, standing in a "submissive" posture. Drake and Linda Sagan explained that most human females *are* smaller than males, and that if both figures had been depicted with arms raised, the plaque's discoverers might mistake that for the normal position. And the female's slightly swayed stance was intended to demonstrate both the mobility of the human hip and movement on the balls of the feet. Rationalizations, cried the critics. The plaque design was obviously yet another example of male chauvinism, even if drawn by a woman who considered herself a feminist.

Then there was the troublesome question of the humans' ethnicity. Linda Sagan's efforts to achieve racial neutrality succeeded but not in the way she'd intended. Rather than objecting that the figures failed to depict their own ethnic group, many criticized the drawing for excluding others. Whites tended to see the figures as white; blacks saw them as black; Asians saw them with Asian features. Some took pride in the belief that their race had been chosen to represent all humanity, while others considered the apparently blatant exclusion of others to be terribly racist.

Perhaps the most bizarre response to the plaque came from a man who insisted the male figure's upraised arm was nothing less than a Nazi salute and wanted to send out another spacecraft to pursue and destroy *Pioneer 10*, presumably to thwart the interstellar spread of fascism.

The Pioneer plaque made a fascinating case study in pop culture psychology. "We didn't realize it, but it turns out to be a cleverly disguised Rorschach inkblot test," Frank Drake mused later. Carl Sagan told *Esquire*'s Cynthia Ozick that "the message is for us too. To remind us that we're on a hurtling mote flung around a mote of fire, and have only what we have." With his characteristic insight, he added: "The human beings represented on the plaque are the most mysterious part of the message."

The plaque was only incidental to Pioneer's true objectives of solar system exploration, really nothing more than an afterthought, a frill, a whimsical extravagance permitted to a couple of fairly influential scientists. Yet this humble aluminum rectangle had come to represent the entire mission in the public eye. Some 20 years later, fervent

protesters would gather at Cape Canaveral to noisily protest plutonium power units on other deep-space probes, concerned that a
launch mishap would spread deadly radiation through the atmosphere. In 1972, though, few people cared about the Pioneer craft's
four RTGs, loaded with plutonium-238. Instead, the public was concerned, dismayed, even angered by the idea of sending drawings of
nude human beings into the cosmos.

Sagan and Drake found the strong reaction to the Pioneer plaque
surprising yet encouraging. Obviously, quite a few human beings harbored great interest in the possibility of extraterrestrial intelligences
and how to communicate with them. That alone made it a success, as
far as Sagan and Drake were concerned. "We had already started to
think of it more as a message *to* Earth than as a message *from* Earth,"
Drake wrote in a memoir. "We had hoped it would hammer home the
idea that we are not alone in the universe, that others will learn of our
existence someday, and that some forms of contact and communication are possible." For two dedicated scientists who had spent so much
of their lives trying to convince their colleagues and the public of the
feasibility of such ideas, the response to the plaque was nothing less
than a personal philosophical victory.

Other scientists, including the Pioneer principal investigators,
found the idea of the plaque of only passing concern. "That was a bit
of whimsy, of course," Van Allen opined. Frank McDonald put the
idea of placing the plaque on Pioneer even more succinctly: "Why
not?" he said. To Eric Burgess the plaque was "a cave painting that will
outlast all the caves of Earth."

Pioneer 10 had yet to leave the planet and fulfill the mission for
which it had been built. But it was already surprising people, already
accomplishing much more than anyone expected—if not, perhaps,
in the way anyone had originally intended.

It wouldn't be the first time *Pioneer 10* did just that.

6

Spring at the Cape

Charlie Hall had no time to worry about charges of sending "pornography" into space. He had a spacecraft to launch. The final project readiness meeting was held at Ames on November 10, 1971. Everything was confirmed on track. By Christmas the Atlas-Centaur launch vehicle was at Pad 36A at Cape Kennedy, ready to go. On January 14, 1972, the *Pioneer F* spacecraft left its womb at TRW. Packed into its handy reusable crate and trucked to Long Beach Municipal Airport, it was loaded aboard a bulbous "Mini Guppy" cargo plane and flown directly to Cape Kennedy. The craft was complete, fitted with all of its scientific instruments; only the radioisotope thermoelectric generators (RTGs) remained to be installed. Upon arrival, *Pioneer F* was settled into a special checkout area, and the final testing and verification procedures began.

As engineers and technicians spent the last weeks before the opening of the launch window busily hovering around *Pioneer F* and doting on its every nuance, Hall and his team were acutely aware that the mission had taken on an almost unbearable significance in some quarters. For a few quixotic years the prospect of realizing the Grand Tour had been tantalizingly, maddeningly within reach for NASA and the scientific community; studies had been done, funding had been set aside, spacecraft had even been designed. But just as 1971 ended, Congress killed the Grand Tour, choosing to put more money into the development of the Space Shuttle. The magnificent scheme of sending four highly sophisticated probes, two to Jupiter, Uranus, and Neptune and two to Jupiter, Saturn, and Pluto, thus exploring the entire outer solar system, was dead. Although Congress later threw NASA a bone by approving a greatly scaled down mission called Mariner Jupiter/Saturn (later to be renamed Voyager), NASA administrator James Fletcher told an interviewer that he was "very disappointed we are not able to go ahead with the Grand Tour. . . . It would have captured the imagination of Americans."

With the demise of the Grand Tour, the *New York Times*'s John Noble Wilford observed that the success of Pioneer "becomes even more crucial." Earlier, some had thought of Pioneer as little more than a warmup for the main event. Now, it would be the only look at the outer solar system that scientists would get for years. Instead of the opening act for the Grand Tour extravaganza, it had become the only show in town, at least until the Jet Propulsion Laboratory could regroup and get Voyager on the payroll.

In mid-February, *Pioneer F* was gingerly moved to the gantry at Launch Complex 36A and mated to the Atlas-Centaur launch vehicle, its three booms folded up, its base affixed to a "spin table" that would set the craft spinning upon its release from the third stage. Spacecraft manager Ralph Holtzclaw and his assistants from Ames and TRW performed the final checkout of the craft and its instruments while crawling around inside the cramped quarters of the Atlas-Centaur nose cone, painfully aware that any slip at this point, such as the punctured telescope windows that had earlier plagued Simpson's instrument package, would not only cost money but, even worse, would inevitably delay the launch.

The computation of launch windows is perhaps as elegant and

perfect an example of applied mathematics as ever existed in any text-book. "People on the street would think it'd be complex, but it's rather simple," said Dave Lozier. "In general, it's pretty much an ellipse and a hyperbola that get you there."

The concept is quite basic when you realize that everything in space is moving constantly. On Earth you can see a mountain on the horizon and walk toward it, confident that it will still be there when you arrive. Space travel doesn't quite work that way. An astronaut in Earth orbit, for example, can't get to the Moon simply by pointing the nose of his ship that way and hitting the rockets. By the time he traversed the intervening 238,000 odd miles, he would find himself in deep space, nowhere near the Moon, which would have drifted placidly in its orbit thousands of miles from him during the course of his journey. If he's very lucky, he would still have enough fuel to get back home, but only if this time he made sure to intercept the Earth not where he saw it then but where it would be when he reached it 3 days later. Fortunately, the Earth and Moon are close enough to each other, and the Moon's orbit about the Earth relatively short, so that our astronaut might actually have a shot at getting home.

For interplanetary travel the principles are precisely the same, but the practicalities are more complicated. For a journey from Earth to Jupiter, one could theoretically launch at any time, if the energy and time expended weren't a concern. But energy and time are always of prime importance in spaceflight because both are strictly limited. Rarely, if ever, is there a compelling reason to waste either, even on a relatively short voyage, say from the Earth to the Moon. Other factors figure into the calculation as well: the more massive the craft, the more energy is needed to send it on its way. To a planet hundreds of millions of miles away, the energy—that is, the fuel—required for a random launch might bankrupt the U.S. Treasury, and those who launched such a foolish mission might be long dead before the space-craft ever arrived.

Instead, trajectory specialists calculate the relative positions of Earth and the spacecraft's destination planet in their orbits, determining the period during which a craft launched from Earth can reach the target planet with the minimum expenditure of energy and time. This is the launch window. For *Pioneer F*, that window opened on February 27, 1972, and closed on March 20. If *Pioneer F* couldn't be launched sometime within that period for whatever reason, NASA would have to wait until the next Jupiter launch window opened about 13 months later—when *Pioneer G* was scheduled for launch. Launch windows could be calculated and planned for but not controlled by the desires or needs of human beings. Dave Lozier recalled a telling illustration of this reality. While waiting at Ames during the countdown for the *Pioneer 7* launch, an engineer remarked to him, "Look what we're doing. After all this preparation, we're just sitting here waiting for the Earth to rotate into the proper position."

The final preparatory step before launch was the installation of the RTGs, done only after every other test and adjustment had been made inside the booster shroud. It was a task of rather hair-raising delicacy. Lozier recalled that the RTGs "were kept in a safe facility with limited access. They're hot, too, hundreds of degrees—you can't touch 'em. I saw them through the window of the facility, with everybody wearing radiation badges." When the time came for installation, Lozier got a call from Ralph Holtzclaw, the spacecraft manager: "He said, 'We're gonna put the RTGs on. Why don't you come over and we'll go out to the gantry?' I said, 'I'm still having kids!' I didn't want to go anywhere close to 'em."

Pioneer's launch was scheduled for the evening hours, specifically at 8:52 P.M., because at that time the spacecraft would receive a free extra boost from the Earth's rotation, added to Earth's orbital motion around the Sun. Illuminated by brilliant floodlights, the glistening white Atlas-Centaur was a striking sight against the Florida night and the black Atlantic Ocean beyond the pad. Like all night launches, the departure of Pioneer promised to be a spectacular event.

Some distinguished guests were at the Cape on that evening of Sunday, February 27; aside from NASA administrator James Fletcher, the head of the Atomic Energy Commission (and later secretary of defense) James Schlesinger was on hand to watch *Pioneer F* leave the planet. Fletcher had invited Schlesinger to thank him for his part in securing the RTGs for Pioneer's use. But they, as well as the several hundred other people who had gathered at the Cape to witness the launch, were doomed to disappointment.

First a squall line hit the Cape 2 hours before launch time with lightning and wind gusts up to 46 miles an hour, which forced the pad crew to wheel a protective gantry around the Atlas. After the squalls passed, a power failure in the blockhouse, where the launch controllers plied their trade, delayed the countdown. Then the weather made trouble again, although not at ground level this time: measurements from weather balloons indicated unusually strong high-altitude winds. The wind shear could tear apart the Atlas as it rocketed through the upper atmosphere, resulting in the loss of the *Pioneer F* spacecraft, all its scientific instruments, and the mission itself. Charlie Hall was taking no chances. He scrubbed the launch. Even if the winds had died down, technicians needed to recheck all the Pioneer and Atlas systems to ensure that the power failure hadn't caused any problems. The team stood down for 24 hours, the launch rescheduled for 8:45 Monday night. Much to his dismay, Schlesinger couldn't stick around.

Monday night saw no unexpected power failures, but the winds again delayed things. At speeds of about 115 miles per hour at 45,000 feet, they were again beyond the safe limits of the Atlas vehicle. Hall again postponed the launch, this time until Wednesday, because the Cape was reserved for another launch on Tuesday. Maybe the third time would be the charm.

It wasn't. Instead, Wednesday night proved a carbon copy of the first two attempts, with high-altitude winds keeping *Pioneer F* grounded. Delays were routine in the launch business—no one wants to risk a multimillion-dollar mission without being absolutely sure that everything is as close to perfection as possible—but Pioneer was beginning to look a little jinxed at this point.

Thursday, March 2. Launch was set for 8:25 P.M., and for once

After some frustrating delays, *Pioneer 10* leaves Earth forever on the evening of March 2, 1972.

the winds seemed to be behaving themselves. One more minor technical glitch held up the launch until 8:49, when the Atlas-Centaur finally lit up the night sky in defiance of distant lightning, the roar of its engines drowning out the tentative rumbles of thunder and the applause of the Pioneer team.

The Atlas first stage burned out and dropped away after lifting *Pioneer F* to an altitude of 85 miles and a speed of almost 8,000 miles per hour. Then the Centaur second stage immediately ignited and burned for 7 $^1/_2$ minutes, accelerating *Pioneer F* to nearly 23,000 miles per hour. With the craft well beyond the atmosphere of Earth by now, the nose cone encasing *Pioneer F* opened and fell away. *Pioneer F* was in space at last and had earned itself the name *Pioneer 10*.

The Centaur burned out, and small thrusters fired around the spin table on which the spacecraft was mounted, setting *Pioneer 10* spinning at 60 times a minute. The Centaur fell away and the Thiokol TE-364-4 third stage kicked in, hurtling the spacecraft still faster into the void. Then it burned out and separated. *Pioneer 10* was on its own. The two RTG booms unfolded to their full length, after a brief but unfounded worry that one had failed to deploy, followed by the 17-foot-long magnetometer boom. Just as a spinning ice skater slows her rotation by extending her arms, the deployment of Pioneer's booms slowed the craft's spin to about five times a minute. The attitude control system turned Pioneer around so that the big main antenna dish was pointed toward Earth. The spacecraft would now always look homeward, even as it continued to speed forever farther and farther away.

Seventeen minutes after leaving the pad, *Pioneer 10* was on its way, never to return, moving faster than anything ever made by human beings: 32,114 miles per hour. Eleven hours later, it would cross the orbit of the Moon, tearing across a distance that took the Apollo astronauts 3 days to traverse (although the Moon was nowhere near Pioneer at the time). If it survived the unprecedented journey, *Pioneer 10* would reach Jupiter 22 months later.

For some their part of the mission was essentially over, or changed, now that Pioneer was on its way. Dave Lozier's job, for example, shifted from launch preparations to trajectory computations. TRW's Pioneer project leader Bernard O'Brien and the rest of his crew could relax: "Once we launched the spacecraft, we had no further contractual responsibility." But technicalities of legal responsi-

bility aside, O'Brien, Lassen, and the TRW team were both too con-
scientious and too emotionally involved with the spacecraft they had
created to forget about it simply because the launch was over and
Pioneer 10 was now out of their hands. They would be watching the
spacecraft closely and breathlessly throughout every milestone, up to
and including the planetary encounter.

By the next day *Pioneer 10* was pronounced "happy as a tick on a
sheep" by a JPL spokesman. JPL was handling the tracking and operations of the spacecraft for the immediate postlaunch period, although the command center would shortly shift back upstate to
Pioneer's home base at Ames.

A minor but rather annoying problem soon cropped up with *Pioneer 10*'s stellar reference assembly or as it was less formally known,
its star sensor. This small device was fixed on Canopus, one of the
brightest stars in the sky, to provide a reference for orienting the
spacecraft. But conflicting data from the sensor made it unclear
whether it was fixed onto Canopus, another star, or even a bright
reflection of sunlight from another part of the spacecraft. Another
possibility was simply that the device was still too hot and would
settle down as the spacecraft moved farther from the Sun. In any case,
it was hardly a mission-critical issue. The star sensor wasn't a vital
piece of scientific equipment, and besides, *Pioneer 10* had two other
Sun sensors to use in order to keep its bearings. For a technical glitch,
this one was about as benign as one could hope for.

As *Pioneer 10* settled into its cruise phase after the stresses
of launch, routine adjustments and checkouts ensued. The Atlas-
Centaur booster had placed the spacecraft on a nearly flawless trajectory, so only some minor tweaks to its course were made, one to
ensure that the Jupiter flyby would happen at an optimal time and
another to send the craft behind Io during the encounter in an attempt to pick up some data on that Jovian satellite. As with all spacecraft, *Pioneer 10* was navigated and tracked by analyzing the Doppler
variations in its radio signals—the shifts in wavelength and frequency
transmitted to and from the spacecraft. Computing the difference

between the known values and what's actually received provides data on the craft's velocity and position. The basic principle is exactly the same as computing the distance of a lightning strike by the delay in hearing the accompanying thunderclap: you see the lightning, start counting seconds, and when you hear the thunder, divide by five to find the distance in miles.

Meanwhile, the science instruments were turned on and checked one by one. The particles and fields instruments, such as Van Allen's Geiger tube telescope and Simpson's charged particle instrument, took base readings as *Pioneer 10* passed through the Van Allen radiation belts, providing a calibration for their later work at Jupiter. The optics of some of the other instruments were too sensitive to strong sunlight this close to the Sun, so they were either left inactivated or shielded from light and heat by a slight adjustment in the spacecraft's orientation. By 10 days after *Pioneer 10* left Earth, all its instruments were up and running perfectly, although some of them wouldn't have much to do until reaching Jupiter.

Not that *Pioneer 10* would be completely idle along the way, however. Just days after leaving Earth, Gehrels's imaging photo polarimeter (IPP) was taking measurements of a phenomenon that until then had only been imperfectly observed from the surface of the Earth: the zodiacal light and its counterpart, called the Gegenschein (from the German for "counterglow"). A faint glow on the horizon visible only in very dark skies, the zodiacal light is caused by the scattering of dust particles in the ecliptic, the plane of the solar system. The Gegenschein is a similar glow seen in a direction directly opposite the Sun's position in the sky, thus visible only at midnight (and even then very rarely because the utmost darkness is vital to its observation). Some theories held that the Gegenschein might have some connection with the Earth itself, perhaps associated with the planet's shadow or dust or gas in some sort of "tail" streaming from the Earth. *Pioneer 10*'s detection of the Gegenschein well past Earth orbit, the glow observed in deep space opposite the Sun, proved otherwise; it was definitely of interplanetary, not Earthly, origin. The IPP also provided new data on starlight that was impossible to obtain inside Earth's atmosphere.

As much as it would later prove to be one of the stars of the mission, Gehrels's baby also caused a few anxious moments. One day as *Pioneer 10* cruised peacefully on course to Jupiter, a fishing trawler in the Atlantic Ocean accidentally snagged and broke a transatlantic cable, severing the connection between the Pioneer control center at Ames and a Deep Space Network (DSN) station in South Africa. A major mishap, to be sure, but on this occasion it was potentially disastrous. The IPP had been in the process of making zodiacal light observations of the sky, scanning the faint light in the dark reaches of deep space. But the instrument had to be turned off before its steady scanning brought its field of view too close to the Sun. Direct, unfiltered sunlight pouring into the IPP would completely fry its delicate phototube, rendering *Pioneer 10* completely blind. The mission wouldn't be compromised as far as the particles and fields scientists were concerned, but there would be no fabulous pictures of Jupiter or anything else for the other scientists, not to mention the press and the public.

For once the admirable simplicity of the Pioneer design worked against it. Without a full-blown computer onboard to handle such contingencies, the instruments had to be controlled from the ground. If communications were disrupted, no commands could be sent. This particular disruption occurred just before the IPP had to be shut off. Somehow, the shutdown command had to get to the South African DSN station—the only station in touch with *Pioneer 10* at the time— and relayed up to the spacecraft.

Pioneer controllers first tried to reroute communications through a satellite, only to find that the satellite's ground station was also out of commission thanks to the snagged cable. With literally minutes to spare, an alternate teletype line was secured through England, across Europe, and finally down to Johannesburg, and the IPP was closed down just before it would have been totally blinded.

The episode was a harrowing demonstration of the truth that, while unmanned missions might not seem as exciting as manned flights to the uninitiated, they could offer more than their share of cliffhangers and close scrapes. Even though this one had been a threat more to one part of the mission than to the spacecraft itself, the Pioneer team had pulled it out in the proverbial nick of time. It wouldn't be the last brush with disaster they would face.

The IPP had another little trick to play. Gehrels received a call in Arizona one afternoon from Ames reporting that his instrument seemed to show an object, apparently a previously unknown comet, directly in front of *Pioneer 10*'s flight path and obviously on a collision course. "I caught a flight to San Francisco, and we agonized for hours that night," Gehrels recalled in his memoirs. "At the Palomar Schmidt Telescope I had seen 'comets' that were actually caused by stray light from Jupiter or a bright star reflected by some bolt or part of the telescope at just the right angle." Was this the same innocuous phenomenon or a celestial projectile homing in on *Pioneer 10* threatening to smash it into oblivion? Finally, the controllers decided to try pointing the spacecraft in a slightly different attitude: sure enough, the "comet" promptly disappeared. It had been only stray sunlight, apparently reflected off the antenna dish into the IPP.

By May, *Pioneer 10* had reached the vicinity of Mars, the playground of the "Great Galactic Ghoul." The Ghoul seemed to be either sleeping or wreaking its mischief elsewhere, however, because the craft sailed past the orbit of Mars completely unmolested on May 24, entering space incognito for the first time. Ahead lay the asteroid belt. Hall, the science team, and the Pioneer designers were confident that the belt would prove no barrier and that the spacecraft and mission would prevail without incident. "We feel that the odds of being hit are about the same as for someone driving across the United States being hit by a 1947 Hudson," Bernard O'Brien waggishly remarked. "But it could happen."

Still, there were a few nagging doubts. Two instruments aboard *Pioneer 10* specifically designed to study interplanetary dust and meteoroids had been registering a surprisingly high level of dust particles on the flight thus far. Was it going to get worse inside the asteroid belt proper? And if so, would the spacecraft be in danger?

BEYOND THE BELT

The test began on July 15, 1972, when *Pioneer 10* entered the asteroid belt. Now attention focused on two instruments: Robert Soberman's "Sisyphus" telescope or asteroid-meteoroid detector, and William Kinard's meteoroid detection experiment. These, and to a lesser extent Gehrels's IPP, would tell the story of the asteroid belt for the first time—whether it could be safely traversed by the spacecraft that would follow *Pioneer 10*, or whether it truly was the impenetrable barrier that some still feared.

Each of the instruments took a different approach to the detection problem. The simplest solution was Kinard's. His instrument consisted of 234 small pressurized cells or, as Charlie Hall called them, "air mattresses," filled with a mixture of argon and nitrogen gas. If a particle punctured a cell, the cell lost gas at a rate proportional to the size of the particle, and the data were registered electronically. The pressure cells were mounted in 13 panels of 18 cells each on the back of the spacecraft's main antenna dish, so that they would face in the direction of travel. It was as if someone had covered the front end of their car with balloons to drive through a hailstorm, although Kinard and his fellow principal investigators fervently hoped that *Pioneer 10* wouldn't encounter anything so violent.

Soberman's Sisyphus instrument was rather more traditional. It used four nonimaging telescopes with photomultiplier tubes that could detect sunlight reflected from passing particles or asteroids. The fields of view of the telescopes overlapped slightly so that an object could be detected by more than one scope at the same time and the data compared and combined. Soberman's results, and the conclusions he drew from them, would later prove to be quite controversial. Some, including Soberman, insisted that Sisyphus had made a dis-

covery of fundamental importance, while others discounted the data
as flawed, the experiment as faulty, and Soberman's claims as ludicrous.

For now, though, that was in the future. The immediate issue was
survival. A NASA press release hedged its bets, reporting that, "based
on a variety of analyses, Pioneer Project officials . . . expect the spacecraft to safely complete its seven-month passage through the belt
without serious damage. . . . However, since no spacecraft has ever
entered the belt before, its actual contents are largely unknown."

If the asteroid belt was so potentially dangerous, some reporters
inquired, why not simply avoid it by flying over or under it? A good
question but one that disregards one of the cardinal rules of spaceflight: don't waste energy. The asteroid belt is mainly located in the
ecliptic plane of the solar system, and flying "above" or "below" it is
certainly possible but only with enormously huge and expensive
launch vehicles. And as usual, time is also a factor: it takes longer to
go around than through. Sending a spacecraft outside the ecliptic
plane engenders other difficulties of tracking, communications, and
navigation. So like it or not, given Earthly budget limitations and
technological capabilities, the path to the outer solar system is strewn
with asteroids.

The worry was less about a large rock taking out the entire spacecraft than about a tiny particle penetrating a vital component in just
the wrong way. At *Pioneer 10*'s great velocity, a particle even the size
of a grain of sand could pierce the hull with more energy than a high-
powered rifle bullet. Even a later failure of the craft at Jupiter would
be easier to take because at least in that case *some* valuable data would
have been obtained. But as Hall explained, "If we had a loss of the
spacecraft while we're in the asteroid belt, nothing else can happen."
The mission would be over.

Aside from the scientists whose instruments were intended to
detect asteroids and dust particles, at least one other PI wasn't too
worried. "I had carefully analyzed the antegrade motion of asteroid
particles relative to the spacecraft trajectory," John Simpson recalled.

"I chose my position [of his instrument on the spacecraft] to look out so that I'd be protected from direct impacts. At first people wondered why I was choosing this odd position. Then the lights go on and they realize it's too late, I chose the one spot where it's free from possible damage."

Only time would tell. Bernard O'Brien explained: "The belt was so wide that you couldn't just sit there and hold your breath, so we just sat back and waited." During the wait, *Pioneer 10* continued its observations of the interplanetary medium and the zodiacal light.

And it did something else that hadn't been part of the original mission plan but was just another case of a Pioneer craft being at the right place at the right time. On August 2, 1972, the Sun erupted in the greatest solar storm that had ever been recorded by humans up to that time. Five days later another storm wracked the Sun, releasing in several hours the equivalent of enough energy to power the entire electrical grid of the United States for 100 million years (at least at the 1972 rate of consumption). The storms flooded interplanetary space with radiation in the form of charged particles of solar wind, pouring into the solar system at unheard of velocities.

Once more, the Pioneer solar weather network went into action. *Pioneers 6* through *9* were the first to feel the brunt of the solar eruptions, measuring the highest solar wind speeds ever recorded—about 2.24 million miles per hour. When the storms occurred, *Pioneers 9* and *10* happened to be lined up with each other with respect to the Sun. Yet by the time the solar wind had reached *Pioneer 10*, 132 million miles farther from the Sun, slightly over 3 days later, the velocity of the solar wind had dropped to about half that measured by *Pioneer 9*. But *Pioneer 10* also discovered that what the solar wind gave up in velocity, it gained in temperature, recording a temperature of about 2 million degrees Kelvin (although the density of the solar wind material was far too diffuse to impart such temperatures to the spacecraft). It was the first time scientists had enjoyed the chance to sample the solar wind, and such huge solar storms, across such a wide swath of the solar system simultaneously. It would be far from the first time that *Pioneer 10* would track the solar wind and observe the convolutions of the Sun.

After the dire predictions of disaster, the asteroid belt had turned out to be something of an anticlimax. "There's nothing surprising to date," Charlie Hall remarked to reporters over the summer. "I suppose the surprising thing is that there has been nothing surprising." *Pioneer 10* passed no closer than 5.5 million miles to any known asteroids, and nothing alarming passed any closer to the spacecraft. Even Soberman's and Kinard's instruments registered much less of the smaller debris than had been expected. On February 15, 1973, *Pioneer 10* was declared to have left the asteroid belt, none the worse for wear. The path was open to the rest of the solar system. To mark the occasion, Ames Research Center held a press conference to announce some of *Pioneer 10*'s preliminary scientific findings regarding the solar wind, zodiacal light, and asteroid belt.

Meanwhile, *Pioneer 10*'s twin sister, *Pioneer G*, was delivered to the Cape and readied for launch. *Pioneer G* was identical to its older sister (including the plaque) with one exception—an additional magnetometer to provide better measurements of the Jovian magnetic fields. If *Pioneer 10* met with unexpected catastrophe during the rest of its mission, *Pioneer G* would complete the mission objectives; if *Pioneer 10* succeeded, its sister ship would fly a different course past Jupiter, checking out details that *Pioneer 10* had missed.

Pioneer G left Earth to become *Pioneer 11* on the evening of April 5, 1973, after a slight launch delay because of stormy weather at the Cape. Unlike *Pioneer 10*, however, *Pioneer 11* gave Hall and his team a bit of a scare. When the command to deploy the RTG booms was given, only one deployed fully. The other seemed to be stuck. The same thing had happened with *Pioneer 10*, but in that case the apparently stuck RTG boom had proved to be a false alarm. Not this time.

Without the booms fully extended, the spacecraft would continue to spin too fast. Unless the stubborn RTG boom could be extended, the mission might already be over. After hurried consultations and a few tense hours, controllers tried repeatedly firing the spacecraft's thrusters, hoping to jar the boom into place. That helped, but not enough—the boom still wasn't properly and fully deployed. Controllers were stumped until fortune smiled on Pioneer once again. When the spacecraft was reoriented in a routine maneuver that would prevent excessive solar heating of the science instruments, the RTG boom abruptly extended to its full length. Now *Pioneer 11* could settle into cruise mode, following *Pioneer 10* to Jupiter. It would arrive about a year later.

Millions of miles ahead *Pioneer 10* continued to hurtle through unmapped space, right on course for its encounter with a world no one had ever seen close up.

7

Twelve Generations from Galileo

I t was such a tiny thing, so small and spindly, that its mere existence was a challenge both to the infinite abyss through which it
was moving and to the immense planet that was its destination.
As *Pioneer 10* hurtled forward, beginning to accelerate as Jupiter's
gravity reached out and began pulling it in, it kept its main antenna
dish always pointed toward Earth, like an insecure runaway who
knows she's left forever and is never going back but can't yet bring
herself to cut the tenuous contact with home. Spinning at the stately
rate of about five revolutions per minute, the spacecraft responded to
every query and every command from Earth with the utmost obedi-
ence, even as its movement inexorably stretched out the radio waves
that carried that vital information. Someday, perhaps, that invisible
electromagnetic link would be stretched so far by distance that it
would break forever, but that eventuality was still far in the future.
For now, *Pioneer 10* was in firm contact with the home planet as it
approached its mission objective.

Whether it would survive the approach was still an open question. The asteroid belt had proved to be no obstacle, with the

spacecraft's instruments registering only a few microscopic and benign impacts with rocky particles. But the surprising sparseness of the belt had some scientists worried that perhaps *Pioneer 10* owed its safe passage to Jupiter's overwhelming gravity. Maybe the massive planet pulled much of the dust and debris of the asteroid belt toward itself. If that was the case, then instead of encountering fewer particles as it drew closer to Jupiter, the spacecraft might actually begin to find more and more, perhaps enough to pose the same threat of fatal impact that some had thought likely in the asteroid belt. Maybe Jupiter was surrounded by a cloud of planetary refuse waiting to destroy any speeding spacecraft that dared to approach.

But that was still only dire speculation by people whose job it was to imagine and plan for all the possible worst-case scenarios. Jupiter's intense radiation was a known, somber fact. Just how strong it might be and of what precise nature were still unknowns, chief among the many questions *Pioneer 10* had been sent to answer. But the most important question was whether the little spacecraft could even withstand the onslaught of Jovian radiation at all. Charged particle radiation—a strong flux of protons or electrons—could create short circuits in *Pioneer 10*'s electronics, causing it to execute spurious commands; or it could simply fry the sensitive solid-state components. The same sort of disasters could ensue if the planet's radiation or magnetic fields created charge buildups in the spacecraft's circuits that resulted in electrical arcing or errant sparks between components.

Pioneer 10's builders and the scientists who had constructed the instruments had done their best to anticipate such mishaps, but there were limits to what they could do. They were designing for an environment whose characteristics were largely unknown, after all, and complete safety was a chimerical goal in any case. As Charlie Hall told reporters when asked about the dangers of radiation and impacting meteorites: "Had we used the more pessimistic estimates, we would have probably had to add 65 to 100 pounds of shielding. The scientific payload is only 65 pounds, so it would have looked foolish to have no instruments onboard the spacecraft to measure the hazard that we were protecting. So we took the other view and went to a more optimistic approach, and primarily used the structure which was required there to support the spacecraft to act as its own shielding." William Kinard, the creator of the "air mattresses" that registered dust impacts, echoed Hall's comments in a briefing at New York's Hayden Planetarium. "If we had provided all the protection some people said we would need, we would have had no instrumentation on the spacecraft, just armor plate," he said.

Such comments evoked chuckles from reporters, but their bemused reactions didn't mean the dangers weren't real. Still, there was little anyone on Earth could do now but wait and hope for the best. None of them—not Charlie Hall, not Bernard O'Brien, not any of the scientists, nor anyone else involved with the project—had any illusions. Everyone *hoped* that *Pioneer 10* would emerge from its Jupiter encounter unscathed, but they knew that the danger of fatal catastrophe was part and parcel of exploring a place for the first time. Even if *Pioneer 10* didn't survive, its very demise would provide vital information for the spacecraft that would follow—first and foremost, Jupiter's next visitor, *Pioneer 11*. If *Pioneer 10*'s course proved fatal, there was still plenty of time to aim *Pioneer 11* on a safer trajectory.

And Hall and his team had more things to keep themselves occupied before encounter than just worrying about disasters they couldn't control. They had to prepare to orchestrate the most complex and most distant planetary encounter ever attempted, a task made even more daunting by Pioneer's "remote control" philosophy. The Deep Space Network would have to handle all of the data and commands passing between Earth and the spacecraft while compensating for all the complications caused by the immense distance: not only the over 90-minute round-trip communications delay between Earth and Jupiter dictated by light speed, but also the weakness of the spacecraft's signal, the Doppler shift in frequency and wavelength caused by the distance and *Pioneer 10*'s swift motion, and the data rate limitations imposed by all of these factors.

Anyone who's been on the Internet for several years and witnessed the steady acceleration of connection speeds, from 2,400 baud modems on conventional telephone lines to DSL and beyond, has an intimate appreciation of the importance of data rates. A Web page heavy with graphics and animation might take minutes to download onto a computer with a dial-up connection, while the same page appears instantly with a DSL line. The factors of distance, signal bandwidth and strength, and Doppler shifts affect the transmission of data between a spacecraft and Earth just as surely as the difference between a dial-up modem and a fiber optic DSL line. And to make the problem even more complicated, there's also a time factor. Waiting for a Web page to load with a slow Internet connection is merely an annoyance; it takes longer, but you'll still get the whole thing. But a spacecraft flyby of a planet is over in a couple of hours at most. During that time, all of the data have to be collected and transmitted back to Earth or stored in the spacecraft for later transmission. Otherwise, the data are lost forever.

Since the *Pioneer 10* spacecraft had nothing more than a very limited data storage capacity, practically everything it had to tell Earth about Jupiter during its swift flyby had to be transmitted in real time, as quickly as possible so that the greatest amount of data could be sent. That meant the communications, commands, computations, everything, had to be operating at absolute peak capacity during the encounter period. There would be no second chance at any observations; whatever *Pioneer 10* missed seeing the first time around, it would never see again.

On November 6, 1973, with the closest encounter still a little less than a month away, *Pioneer 10* used Gehrels's imaging photo polarimeter (IPP) to take the first pictures of Jupiter ever obtained from outer space. They were actually rather disappointing, only perhaps slightly better than the best photos taken from the surface of the Earth, but it was still an interplanetary Rubicon, marking that the encounter phase of the mission had begun. The next important milestone would be the crossing of Jupiter's bow shock, the zone in which

Tom Gehrels' imaging photo polarimeter views Jupiter from 1,842,451 miles as *Pioneer 10* closes for encounter.

the solar wind is deflected around the planet by its magnetic field, similar to water rushing past a rock in a creek. The bow shock would be the point at which Pioneer 10 would pass into Jupiter's cosmic backyard.

Much to the surprise of the principal investigators, Jupiter's yard turned out to be much bigger than expected. *Pioneer 10* crossed the planet's bow shock on November 26, days earlier than predicted, at a distance of 108.9 R_j (Jupiter radii), or approximately 4 million miles. Then the spacecraft crossed the magnetopause (the boundary between the influence of the solar wind and the planet's own magnetic fields) and entered Jupiter's magnetosphere proper at about 96.4 R_j. Obviously the magnetic field of the planet stretched out into space much farther than anyone had thought. More strangely, *Pioneer 10* told scientists that Jupiter's magnetic field was inverted—the magnetic north pole was at the planet's south pole, and vice versa. And while at Earth only the magnetosphere serves to deflect the particles of the solar wind, Jupiter has an additional barrier of plasma just inside the magnetosphere boundary. Also odd was the shape of Jupiter's magnetic field. Earth's magnetosphere tends to be teardrop shaped. The round edge points toward the Sun and constitutes the bow shock, while in the direction opposite the Sun the field trails out into a tail as the solar wind whips around and past the disturbance caused by Earth's presence. But Jupiter's magnetosphere seemed to be flatter, shaped much like a plate or disc with the planet at the center. Even odder, the magnetic field wobbled up and down like a plate spinning on a stick.

These findings were only the tantalizing beginnings of what Van Allen calls "a period of intense discovery." The quiet part of the mission was over. The slow, uneventful interplanetary drift, the tension of the asteroid belt transit, were in the past. *Pioneer 10* was about to fulfill the purpose of its existence, whether it survived or ended its life in a blaze of glory. For the scientists, even though the holiday was still weeks away on the calendar, it was the night before Christmas. As the data began to pile up, the PIs reviewed it on the spot, and sent back precious copies for more detailed analysis back at their home institutions. Frank McDonald remembered: "We would get the data tapes from Ames, take 'em up to the San Francisco airport, and give them to the pilot or made special arrangements to fly them back to Goddard and play them through our system at Goddard and send back plots."

The pace of events was picking up rapidly with every mile that closed between *Pioneer 10* and Jupiter. Most of the PIs and their teams of colleagues and graduate students were already on the premises, some of them already basking in an increasing flow of data even before *Pioneer 10*'s closest approach to Jupiter. Meanwhile, the press began to descend on Ames Research Center, and although previously considered a NASA backwater unaccustomed to the spotlight, the center adapted quickly to the influx of reporters and media attention.

"I think most of the center personnel were astounded at the number of reporters and television crews that descended on their relatively small facility," said Robert Kraemer. "It was really great for morale, even for those just tightening bolts in the Ames wind tunnels and not connected in any direct way with Pioneer." No longer languishing as perhaps NASA's most obscure outpost, Ames had been promoted almost overnight to the center of attention from the world's scientific press. After years of being overshadowed by more glamorous NASA centers such as JPL and Houston's Manned Spaceflight Center, Ames had finally found its place in the Sun.

"The news media interest was so keen," Richard Fimmel remembered. "I once asked some of the reporters, from overseas, Japan, all over the world, 'What are you guys doing here? This is just a little spacecraft. I'm surprised.' The answer I got almost invariably, at least from the U.S. news media, is that there's so much bad news in the world, that if there's some good news, and we think this is, we want to get it. They just descended on us in droves." The enthusiasm of the media was made especially obvious to Fimmel one evening during the encounter period when his quiet dinner at home was interrupted by a TV news van pulling up in his driveway. Reporters collared him in his living room for a live on-air interview.

As project science chief, one of Fimmel's responsibilities was keeping tabs on the health of *Pioneer 10*'s scientific instruments, including the one that was drawing the most public attention, Gehrels's IPP. The images were getting better and better, promising even more spectacular pictures by the time of *Pioneer*'s closest approach to Jupiter. "As we were going to Jupiter there was an intensive amount of work with the IPP," said Fimmel. "We coded over 10,000 commands on punchcards. I would be traveling back and forth to the University of Arizona and working with the team and grad students. I'd get lunch brought in to me and work right through the day, because if I had a flight coming back that day I wanted to get as much done as we could."

As *Pioneer 10* spun about its axis, the IPP scanned the planet in progressive strips, each slightly overlapping the last, in red and blue light. Never-before-seen details of the Jovian cloud formations and the Great Red Spot were making themselves visible. L. Ralph Baker of the University of Arizona had developed a way to view the IPP images in real time on a television screen called the PICS (Pioneer Image Converter System). Aside from providing a convenient check on the IPP's proper operation, the PICS made it possible for reporters and other interested parties to watch pictures of Jupiter being built up piece by piece on TV monitors in the main auditorium at Ames. To make the pictures clearer and more aesthetically pleasing, the PICS was also capable of adding color to the images, since the unprocessed red and blue light pictures were of value only to the scientists.

The task of colorizing the images fell to operations manager Fred Wirth. He found it to be something of a thankless job. "I'm sitting there with a tiny little knob on the console adjusting the green. I kept looking at it, standing back, saying 'Is that about right?' We did this for the public. But one television camera focused on me as I was adjusting it, and the guy said, 'Are you manufacturing this?!' I mean, they got all excited that I was making up these images."

The PICS images could also, of course, be fed to the TV networks. For the public and the nonscientific quarters of the press, to whom streams of data bits on magnetic fields and particle flux were incomprehensible and tedious, the visual experience provided by the IPP coupled with the PICS brought home the importance of the encounter in a visceral way. It was one thing to hear James Van Allen wax poetic about relativistic electrons and synchrotron radiation, but it was quite another to actually watch pictures of Jupiter being slowly and steadily pieced together before your own eyes on a television screen. It was enough to earn Ames a singular honor: an Emmy award from the National Academy of Television Arts and Sciences.

Ironically, the PICS imaging system was almost a casualty of Charlie Hall's passion for practicality. "I needed an extra 10 or 20,000 dollars so that we could implement in the tape processing station a system for real-time image production as we were going by Jupiter [the PICS]," Richard Fimmel remembered. "And Charlie said, 'Who cares about that? So we'll get the pictures a day or two later.' That was his attitude. He didn't feel it was that important. I thought it was because of the interest people had. We went back and forth, and finally he said, 'All right, Rich, you can have it.' Later we're going by Jupiter, and here's Charlie in the control room with me, watching the image build up scan by scan, and he says, 'Gee, Rich, this is great!' I said, 'Yeah, Charlie, I remember telling you and you didn't think it was such a good idea!' He said, 'Yeah, you were right and I was wrong!'" Fimmel laughed. "It was the only time in my life that I knew Charlie said someone else was right and he was wrong."

As Eric Burgess remembered, Charlie Hall had never even seen Jupiter through a telescope before, which caused some amusement among the Pioneer team: "I pulled his leg about that one," he said. But even for the supposedly more jaded types such as Hall, the visual element emphasized the spectacular nature of what was about to happen. Yes, spacecraft had flown by Venus and Mars before and transmitted amazing sights back to Earth. Still, those worlds were closer to our own, and thus more familiar and less alien somehow. Jupiter was the undisputed giant of the solar system, more distant and more mysterious and yet more compelling in many ways than our planetary neighbors. Some scientists had rather archly observed that the solar system consists of the Sun, Jupiter, and some debris. We could see the Sun every day; we had already examined some of the "debris," including our own little speck of it; now we were seeing the other major denizen of the solar system closer than ever before.

Indeed, if you stopped to think about it, the whole enterprise seemed almost arrogant. Sending a 570-pound spacecraft to the solar system's largest planet, a world so huge that 11 Earths could fit across its radius, was so audacious that it was difficult to find words to describe the endeavor's magnitude and all too easy to unintentionally downplay it. Hans Mark, the director of Ames, certainly displayed a mastery of understatement when he told newspeople: "This is an unusual event." But at the same press conference, Charlie Hall did a better job at conveying the significance of *Pioneer 10*. "We are really only 12 generations away from Galileo and his first crude look at the planet," he said. "Twelve generations later, we are actually there measuring many of the characteristics of the planet itself." Galileo himself would no doubt have been pleased and utterly astounded at how far humanity had come since he first observed Jupiter through his small telescope.

The PI teams, and just about everyone else, were on a round-the-clock schedule by now. "All the PIs were assigned small cubicles in a big hangar at Ames," Van Allen remembered. "And we'd get a runner from the telemetry reception desk carrying these big sheets of computer paper and distributing them about every hour to the investigator groups." Around this time the scientists became intimately familiar with what would become one of Charlie Hall's most famous—or perhaps infamous—management techniques.

Each morning of the encounter period began with a meeting in Hall's small office of the PIs and other project members. But these were not leisurely, relaxed coffee klatsches over donuts and crullers. "He refused to let anyone sit down," John Simpson said, laughing. "You had to stand in a big circle, so your reports were short and to the point." Van Allen added: "That was a very effective technique. Charlie and the rest of us stood through the meeting and gave a 5-minute report on what we found out during the previous 24 hours and showed sketches and graphs."

Tom Gehrels laughed: "We did what he said, and it was amazing, because we were all rather prima donnas."

"The stand-up meetings were an absolute riot," Fred Wirth recalled. "He deliberately kept the room warm. . . . And I'll tell ya, after about a hour and a half or 2 hours in that stand-up meeting, you couldn't stand anymore, you couldn't breathe anymore, and it became almost intolerable. That's what [Hall] did deliberately to keep the meetings short. We had the conference room right down the hall. He could have parked 20 people in the conference room, and they could have gone to the blackboard and done their thing, but no, he

Charlie Hall (center) presides over one of his trademark "stand-up" meetings. At bottom right are the hands of one participant who seems to be cheating by sitting down.

wanted it in his office. He deliberately made the meetings so uncomfortable to make them short. It was part of his strategy."

An effective strategy it was. Hall's stand-up meetings kept things moving, kept everyone informed, and are still universally remembered by all involved with both amusement and perhaps some retroactive backaches. However physically fatiguing those meetings might have been for the participants, the intellectual excitement they created was infectious. One morning Hall came in early to find two scientists from NASA headquarters slumbering by his office door. They had camped out to make sure they would be able to find a place inside Hall's small office for the morning's meeting. Van Allen still tells the story of how Nobel laureate scientist Luis Alvarez, who happened to attend one of Hall's meetings as a special guest, remarked afterward on the "incandescence" of the experience, as awed and excited scientists shared their latest findings with each other and the man who led the entire project.

The admiration was mutual. "We had one hell of a good bunch of scientists," Hall recalled. "They were all gentlemen. They worked together 100 percent of the time. There was no, 'That's my data. You can't have it,' or 'I figured that out first, so you can't use it.' I would watch the scientists and think, 'God, what a team we've got.'"

After the PIs shared their findings with Hall and each other, Hall and project scientist John Wolfe would then pass along the latest to the media in press conferences that were held every few days. The occasions were anything but dry and boring. Hall and Wolfe made such a good team—informative, receptive, and just plain entertaining—that the press quickly dubbed their briefings "The John and Charlie Show" and later gave Hall and Wolfe a plaque with their caricatures and the signatures of the press corps. Sometimes the John and Charlie Show would feature guest appearances by some of the PIs, who would field questions from eager reporters.

Periapsis, *Pioneer 10*'s closest approach to Jupiter, would come on Monday, December 3, at 6:24 P.M. Pacific time. It was actually about a minute earlier than previously calculated by Jack Dyer and his trajectory wizards; Jupiter had turned out to be more massive than believed and thus tugged on *Pioneer 10* with greater gravity than expected. But as the spacecraft sped closer, accelerating up to 82,000 miles per hour, no one was really worrying about the precise time of periapsis. Neither had the idea that Jupiter might be surrounded by perilous dust proven to be a real problem, although *Pioneer 10* had found that the amount of dust around Jupiter was still about 300 times that of open interplanetary space. Instead, it was the radiation level that was making everyone sweat, and not just in the figurative sense.

Jupiter's intense radiation was no surprise, of course. One of the prime objectives of the mission was to learn more about it. But now that reams of hard data on that radiation environment were pouring into Ames every hour, the PIs were able to extrapolate more precisely about just what sort of a lion's den *Pioneer 10* was inexorably penetrating. The predictions, to say the least, were not encouraging. The spacecraft was on course to pass the planet at a distance of only 1.86 Jupiter radii, and although that had been thought a safe enough distance, now it was beginning to look as though it might still be too close. Changing the spacecraft's trajectory was out of the question; it was far too late for the craft's weak thrusters to seriously defy the gravitational grasp of the giant planet. For better or worse, the trajectory die was cast and with it *Pioneer 10*'s fate.

The mood among the PIs grew darker with every new model, every updated prediction. "Misery of miseries, they now predicted a level of radiation about a factor of 10 worse than had been specified for our instruments!" Tom Gehrels wrote in his memoir. The desolate scenario, he confirmed, was clear to everyone: "Our imaging photopolarimeter with its subtle and sensitive optical components would send confused signals at about noon, Pacific time, and soon thereafter die forever. The other instruments would give up similarly later in the afternoon, and by about 4 PM the spacecraft transmitter too would garble out its very last signals." No one back on Earth would witness *Pioneer 10*'s closest approach to Jupiter because "by then the Pioneer spacecraft would be silent, dead."

Hall wasn't quite so pessimistic, figuring that the dire predictions of radiation catastrophe were exaggerated. Hours before encounter, however, he was beginning to change his mind, when he walked into "what looked like a funeral parlor." Scientists were steadily monitoring the Jovian radiation levels. "They were plotting energy versus distance to the planet," Hall said, "and that line was going straight up." If the radiation kept increasing, *Pioneer 10* was doomed. "The TRW en-

Artist's conception of *Pioneer 10* during its closest approach to Jupiter.

gineers said flatly that *Pioneer 10* was about to go off the air," Robert Kraemer wrote. "Everything on the spacecraft was being fried."

"One guy said, 'My god, we've had it,'" Hall recalled. "And not 5 minutes after he said that, the peak was reached and it started to go down. So actually the radiation intensity was less after we passed a certain point on the way in. You'd have thought the peak would have continued up as we got closer to the planet, but that's not what happened." The radiation levels fell off rapidly and, once again, *Pioneer 10* had survived a narrow scrape. "We were all cheering that the little spacecraft had made it," Van Allen said.

The peculiar configuration of Jupiter's magnetosphere had saved *Pioneer 10*. William Dixon of TRW called it a "magnetodisc," and it was an apt description, a flattened disc of magnetism inside of which were trapped the charged particles that kept the Jovian environment so dangerously drenched in radiation. But the magnetodisc "wobbled" about the planet as it spun, tilted with respect to Jupiter's equator, so that it actually moves up and down from one side of the planet to another. Fortunately for *Pioneer 10*, the magnetodisc had wobbled up away from the spacecraft during the planetary approach, carrying away its deadly radiation and clearing the path to periapsis.

To illustrate the concept for reporters, Van Allen called back home to the University of Iowa and had some graduate students hurriedly send some special equipment out to California: a painted sphere of Jupiter, lined with a specific pattern of holes, mounted on a electric motor-driven shaft. They also threw in a generous supply of wire. Van Allen flattened and twisted the lengths of wire, inserting the ends into the holes of the Jupiter sphere to represent the magnetodisc. At the next press conference, his battery-operated model of the Jovian magnetic fields proved a big hit and a crude but revealing illustration.

Not that *Pioneer 10* was completely unscathed. Some images, in-

James Van Allen debuts his Jovian magnetosphere model (foreground) as fellow PI John Simpson looks on.

cluding of Jupiter's moon Io, were lost because of some false commands generated by the radiation in the IPP circuitry. To protect the operation of the IPP from spurious radiation-induced signals, a special sequence of commands had been devised to reconfigure the instrument to compensate for any errors, but as a result some Io images were lost. Considering that the demise of the IPP would have meant a loss of all the close-up images of Jupiter, it was considered a small price to pay. For a brief period, two of the cosmic ray telescopes in Frank McDonald's instrument had been completely saturated by the radiation, but they returned to normal after the levels dropped off.

Only Soberman's Sisyphus instrument was a complete casualty. The Jovian radiation levels totally opacified its optics and destroyed its photomultiplier tubes, rendering it useless for further observations. Soberman was unfazed, however. "Since we thought and everybody thought our experiment was mostly for the asteroid belt, the fact that the electronics would likely fail at Jupiter because of the radiation field was something we accepted. For the same reason, NASA rejected our request when we asked for glass coverings for the photomultipliers that wouldn't darken in the radiation field. It would have been $200 more. Believe it or not, the electronics survived, but the radiation had darkened the photomultipliers, so after Jupiter it was a complete washout." Of course, the loss of his instrument meant that Soberman and his team didn't have much more to do. "I would just stay in the background and keep out of the way of the people who were working."

So *Pioneer 10* had survived the most dire predictions of disaster, with the loss of only one instrument. Not bad, considering that the spacecraft had taken a dose of radiation 1,000 times stronger than what would kill a human being: about 500,000 rads.

Finally *Pioneer 10* did what it had set out to do almost 2 years ago, right on schedule. Just before 6:30 P.M. on December 3, the craft passed through periapsis, 81,000 miles above Jupiter's swirling cloud tops, still very much alive. Then began what may have been the most anxious moments of the entire mission, more than the launch, more than the passage through the asteroid belt: *Pioneer 10's passage be-hind* the giant planet. For just over an hour the spacecraft would be out of touch with home, for the first time since its departure. With a spacecraft as dependent on its communications link with Earth as *Pioneer 10*, that thought was sobering enough. What made it worse was the possibility that after the unavoidable blackout period, Earth might never hear from *Pioneer 10* again because the little craft might have finally succumbed to the planet's still intense radiation. Or contact might be reestablished only to discover that the scientific instruments, particularly the IPP, were finished. "Everybody was scared and praying it would be alive when it emerged," said TRW's Bernard O'Brien, who was at Ames to watch his company's baby during its finest hour.

Sixty-five minutes of standing around, chewing nails and pencils, and gulping the always-present coffee ensued as everyone waited for *Pioneer 10* to call Earth. Few groups of human beings are as anxious or as helpless as a team of scientists, engineers, and flight controllers waiting to hear from an out-of-touch spacecraft. Some resort to gambling. O'Brien admitted: "I made a bet with one of our scientists that we would survive the radiation belt of Jupiter and he bet we wouldn't." Others are more physically active. "I recall charging down the hall pushing people out of the way to get to the control room to get the commands off to reconfigure the instruments," Richard Fimmel said.

Slowly, as the time for signal reacquisition arrived, an image began to appear, pixel by pixel, on the PICS at Ames. The Pioneer team watched a bright line slowly becoming a crescent on their screens. They were seeing something no one had ever witnessed before: sunrise on Jupiter, as the IPP looked back toward the planet. Over the next hours, as *Pioneer 10* began its eternal journey away from Jupiter and out of the solar system, it continued to send back some final postcard pictures of a crescent Jupiter, something impossible to see from Earth.

Hall, O'Brien, and the rest of the Pioneer team had won their bets. *Pioneer 10* accomplished its last major objective by using

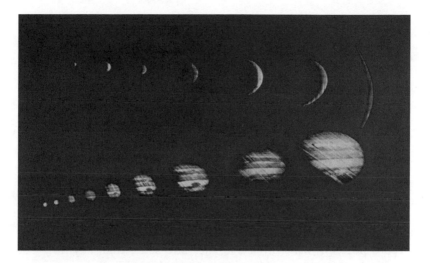

A collage of *Pioneer 10* Jupiter images, including the crescent Jupiter seen as the spacecraft left the planet.

Jupiter's gravity and orbital momentum to hurl itself out of the solar system toward interstellar space at 25,000 miles per hour, proving the feasibility of the gravity boost technique for later missions to Saturn and beyond. And O'Brien claimed his prize: a quart of Beefeater gin.

NASA officials and scientists proclaimed *Pioneer 10*'s mission a complete success and a "beautiful effort," as NASA's Robert Kraemer put it. Charlie Hall told the *New York Times* that it had taken about 25 million man-hours to build *Pioneer 10*, "about the same amount of time spent looking at the first quarter of the Super Bowl. I think the country is getting a lot more out of the spacecraft than the Super Bowl game." In one of the oft-quoted proclamations, Kraemer also told the press, "We can say that we sent *Pioneer 10* off to tweak a dragon's tail, and it did that and more. It gave it a really good yank and it managed to survive." Not only that, *Pioneer 10* mapped the dragon's lair and showed future travelers how to pass through it safely. That was good news for *Pioneer 11*, still making its way through the asteroid belt as its twin sister left Jupiter far behind, and for Voyager, still in the planning stages.

Even President Richard Nixon, like most chief executives who had little use for NASA or space exploration in general, managed some generic words of praise, calling the mission "an impressive technical scientific achievement" and an example of how "man's ability to explore the heavens is on the threshold of the infinite." Given the Watergate scandal in which he was embroiled at the time, he might be forgiven for neglecting to mention that such explorations tend to be quite finite when Washington continually and mercilessly slashes NASA budgets, as his administration had done with both the Grand Tour and the Apollo program. He would not remain in office to praise the achievements of *Pioneer 10*'s successor a year later.

No one could contest the fact that *Pioneer 10* was one for the record books. It had traveled farther into space than anything else ever made by humans and was stretching the distance with every second. It had broken the space speed record, first upon leaving Earth when it exceeded 32,000 miles per hour, and then proceeded to break that record when it swung past Jupiter at 82,000 miles per hour, or almost 23 miles *per second*. Of course, *Pioneer 10* was the first spacecraft to venture beyond Mars, through the asteroid belt, and to Jupiter. It was the first spacecraft to use Jupiter's gravity for navigation, the first to use a primary nuclear energy source, the first to communicate over such immense distances. In short, *Pioneer 10* had been the first to do just about everything that a spacecraft needed to do in exploration of the outer solar system. NASA sought official recognition for these achievements by submitting formal applications to the bodies responsible for verifying flight records, the National Aeronautics Association and the Federation Aeronautique Internationale, for the maximum distance traveled from the Sun, the maximum communications distance, and the duration of the mission.

A more whimsical acknowledgment of the *Pioneer 10* mission arrived in the Ames Public Affairs Office shortly after the encounter. It was a letter from a Commander J. P. Dunning of the Royal Navy, captain of Her Majesty's Ship *Jupiter*. "I was more than a little surprised, on 3rd December, to read in the Newspapers that something called Pioneer X was hoping, that night, to navigate extremely close to us in *Jupiter*," wrote the commander. "This additional collision hazard during our crossing of the already tanker-crowded Oman Sea occasioned me sufficient anxiety for me to instruct my Officers of the Watch to call me should Pioneer approach to within 81,000 miles." Commander Dunning included information and photos of the "life species on board": British sailors in full naval uniform.

The exhausted PIs were mostly too exhausted to appreciate such droll humor. They were physically and emotionally drained, buried in mountains of raw data, and badly in need of sleep, but they were also ecstatic. Van Allen called it "one of the most exciting events of my life." The *Pioneer 10* scientists were already preparing the first papers detailing their preliminary results for the scientific journals, and the analysis and interpretation of the *Pioneer 10* data would keep them busy for years to come. And if that wasn't exciting enough, a year later they would be flooded with another set of new data, courtesy of *Pioneer 11*, with which to compare and contrast their *Pioneer 10* results.

The entire Pioneer team considered the project a complete and unadulterated success but thought that the story of *Pioneer 10* was essentially over. There would be some further observations, mostly from the particles and fields instruments, as the spacecraft set off into infinity, but as far as Hall, the scientists, and the rest of the team were concerned, the main mission of *Pioneer 10* was over.

They were only half right. The Jupiter mission was definitely a historic triumph of science and engineering, but the spacecraft would refuse to go gentle into the night of interstellar space. Even as it continued to move inexorably farther and farther from the planet of its origin, *Pioneer 10* had yet more surprises in store.

8

Filling in the Gaps

While its slightly older sister *Pioneer 10* was making history at Jupiter with the eyes of the world on it, *Pioneer 11* made its uneventful way through the asteroid belt, on course for its own rendezvous with the limelight. Along the way, it repeated and expanded on many of the observations *Pioneer 10* had made the first time through, providing further data on the solar wind, the interplanetary medium, and the zodiacal light.

There had been a few minor problems, aside from the stubborn RTG boom after launch. The heart of the main radio transmitter, a device called a traveling wave tube (TWT) that amplified *Pioneer's* signals to enable them to be detected by the Deep Space Network across the void, had been troublesome enough to cause controllers to switch to the transmitter's backup TWT. It wasn't a mission-threatening glitch at this point, but no one wanted to rely on a touchy transmitter during an encounter with Jupiter. "No further problems are expected," a NASA press release explained reassuringly.

Now that *Pioneer 10*'s mission at Jupiter had been resoundingly successful, though, *Pioneer 11* could afford to take some chances, freed of the necessity to repeat its predecessor's job. The principal investigators (PIs), particularly the particles and fields specialists such as Simpson and Van Allen, wanted more information on the Jovian magnetic fields and the charged particles trapped within them, data they could compare and contrast with what *Pioneer 10* had provided. That meant a much closer approach to the planet and thus a deeper penetration of the radiation belts, but after *Pioneer 10*'s narrow escape from radiation oblivion even at 81,000 miles above the planet, going in even closer was obviously suicidal—at least if *Pioneer 11* followed *Pioneer 10*'s more or less equatorial course, which maximized exposure time in the radiation belts. But if *Pioneer 11* took a different trajectory—for example, an approach to the planet at a much greater angle, swinging from south to north pole and almost straight up around the planet—it would be exposed to the flattened magnetic field and its radiation for a much briefer period, even if periapsis was much closer than that of *Pioneer 10*.

This is the path that was chosen, with periapsis of less than one-third the distance of *Pioneer 10*'s closest flyby. It was still risky, of course, but at least the risk was a more calculated one. Thanks to *Pioneer 10*, the radiation environment of Jupiter space was no longer a completely unknown quantity. Having already tweaked the dragon's tail and survived, the Pioneer team knew that they could tease the dragon once more, a little closer this time, and still make it out alive.

And there was another compelling argument for going in much closer to Jupiter. In the planning stages of the *Pioneer 11* mission, an intriguing possibility had emerged: two planetary encounters for the price of one, using the gravity assist technique. "The JPL people had discovered that if you targeted the thing correctly at Jupiter, you would be lifted on to Saturn," said Charlie Hall. "I frankly was against it at the time because I thought, let's pay attention to *Pioneer 10* at Jupiter and get [those] data correctly rather than flitting from *Pioneer 10* at Jupiter and *Pioneer 11* at Saturn." It was the classic Hall "keep it simple" philosophy. But although he steadfastly resisted unnecessary changes and complications in the missions, Hall was also receptive to having his mind changed with a good argument, especially after *Pio-neer 10*'s triumph at Jupiter.

"We had the results from *Pioneer 10*. There was a big project discussion in which some of us took the lead in advocating going on to Saturn," Van Allen remarked. "Some people were opposed to that and thought we should repeat the *Pioneer 10* encounter trajectory. Others thought we should take a trajectory which would be more appropriate to assessing the risks for the prospective Voyager mission."

Van Allen's comment about Voyager is a not-so-veiled reference to what has often been perceived and portrayed as a competition between Ames Research Center and JPL, its downstate NASA rival. The seriousness of this rivalry and how much of a factor it may have been in the planning and conduct of planetary missions are still subjects of debate among Pioneer veterans and other interested parties. JPL had undoubtedly been the nerve center of NASA's planetary exploration from the beginnings of the agency, and its first director, William Pickering, jealously guarded that position, fighting for JPL's primacy even during the trials of the early 1960s when JPL missions (particularly the Ranger lunar probes) failed so embarrassingly often that NASA considered scrapping JPL's entire effort. JPL's later triumphs at Venus and Mars redeemed its reputation and solidified its supremacy, at least in the eyes of Pickering and subsequent directors. But although JPL would remain NASA's key center for planetary exploration, NASA headquarters was no longer sanguine with letting a single center handle such an ambitious program all on its own. The decision to give the Pioneer solar probes to Ames instead of JPL was one result, and it no doubt rankled some at JPL and elsewhere, even if their displeasure wasn't openly expressed. The fact that Charlie Hall made it look so easy, when at the outset few in NASA outside Ames really believed he could pull it off, only made it worse.

JPL had an overwhelming trump card in dealing with its rivals, however. Along with its planetary exploration role, JPL operated the Deep Space Network, essential for communicating with and controlling all missions beyond the Moon. Any other NASA center that needed the use of the DSN had to work with JPL if it wanted to schedule tracking time, uplink and downlink data, and maneuver its deep-space probes. Still, while such arrangements could be tense, they rarely if ever degenerated into outright contentiousness and obstructionism, at least not after a mission was already approved and in process. These were professionals, after all, with a common goal of ensuring the success of every mission—whomever it might belong to officially.

Charlie Hall recalled a meeting at JPL to set up ground rules for JPL's navigation of *Pioneer 10*. At first it was like a conclave of the most ruthless corporate raiders. "Didn't trust the other guy as far as you could throw him across the street," he laughed. "But in short order, you couldn't tell whether the people we appointed were working with JPL or Ames. You turn the job over to the people who are going to do the work and they find a way to work together."

The rivalry between Ames and JPL, and indeed among NASA centers in general, could certainly become unpleasant at times. But for the most part it was more of a healthy competition urging everyone involved to their best effort rather than cutthroat, take-no-prisoners warfare. A high-visibility project of whatever type meant funding, jobs, and prestige to the center chosen to administer it, and those not chosen are apt to feel a bit slighted. But when the work needed to be done, competition became cooperation.

Which is not to say that political maneuvering and infighting were completely unknown. Fred Wirth recalled a maneuver by JPL management to take the Pioneer project away from Ames. "JPL lobbied, and I remember fighting . . . JPL proved in a great big elaborate white paper to NASA headquarters that they could run the Pioneer missions much cheaper than the Pioneer project office at Ames. I remember going to NASA headquarters with Richard Fimmel for this briefing of JPL's proof they could do it cheaper. And we brought our own budgets. All of JPL's numbers were phony; they pulled them out of the air. I spent weeks and weeks poring over their budgets prior to the meeting and refuted half of them. I got sample costs from other missions of JPL, like the Mariners, and I said look, for Mariner they spent this much, and I'm spending this much. We proved to HQ that we could do it cheaper and guaranteed for less money. So eventually JPL caved in."

As *Pioneer 10* encountered Jupiter and *Pioneer 11* followed, JPL was in the midst of planning the Voyager missions to Jupiter and Saturn. No one could deny that Pioneer had been first to Jupiter, but barring a catastrophic mission failure, that was a foregone conclusion because it had been part of the original mission objectives. But going to Saturn had not been one of Pioneer's stated objectives, and when the possibility arose that *Pioneer 11* might actually reach Saturn years before Voyager, a few feathers were ruffled, according to some, including PI Tom Gehrels. "There was an attempt to prevent Pioneer from doing that [going to Saturn] and to let Voyager be the first to Saturn," he maintained. Immediately after *Pioneer 10*'s successful Jupiter encounter, the *New York Times* reported on the possibility that *Pioneer 11* might be directed on to Saturn, noting that "the primary concern at NASA headquarters in Washington is understood to be political, not technical. If *Pioneer 11* is cleared for a Saturn encounter, would that undercut support for the more ambitious Mariner (later renamed Voyager) spacecraft mission to Jupiter and Saturn, scheduled for launching in 1977?"

Whatever 11th-hour maneuvering may have taken place to accomplish such a change, however, it was too little and too late. Besides, even the most competitive JPL stalwarts realized that Pioneer was performing an invaluable service for Voyager by blazing the trail beforehand. Pioneer's experience with the Jovian radiation belts and trajectory planning, for example, was vital to the success of Voyager, and JPL was well aware of it. The Voyager team was also well aware of the potentially deadly hazard posed to their mission by the Saturnian ring system. Just as no one was quite certain whether a spacecraft could survive Jupiter's radiation until *Pioneer 10*, the prospects of a spacecraft surviving passage anywhere near the rings of Saturn were far from clear. Let *Pioneer 11* take the risk, many privately argued.

The possible sacrifice of such a cheap and simple craft would be worth the salvation of the much more expensive and sophisticated Voyager.

So on April 19, 1974, not long after *Pioneer 11* had emerged from the asteroid belt and confirmed *Pioneer 10*'s findings about the lack of small, high-speed dust particles within the belt, and thus the lack of danger to transiting spacecraft, the craft's trajectory was tweaked to place it on course for its barnstorming pass by Jupiter and subsequent slingshot to Saturn. Because of the alignment of the planets at the time of *Pioneer 11*'s Jupiter flyby, *Pioneer 11* would actually be flung back across the solar system in the general direction of the Sun in order to meet up with Saturn. In so doing, it would also be hurled far above the ecliptic plane of the solar system, becoming the first spacecraft to traverse that region of space. It would take *Pioneer 11* almost 6 years to reach the ringed planet, after its Jupiter encounter in December 1974.

Again, it was a calculated gamble. The spacecraft had only been designed to reach Jupiter, after all, not to be flung across the solar system for an extended journey and a second planetary flyby. The RTG power might give out before *Pioneer 11* ever reached Saturn; instrument failures might render the spacecraft useless; or it might not even survive Jupiter as had *Pioneer 10*. But Hall and his team were used to taking chances by now. This seemed like a good one.

TAPES AND PUNCHCARDS

In an age of microprocessors and handheld computers when digital technology is ubiquitous, it can be difficult to conceive of a time when computers used reels of magnetic tape, filled rooms, and used punchcards and teletypes for interfacing. But Pioneer is a product of just such a time, and computer technology was both an indispensable aspect of the project and a source of problems.

"Charlie Hall was dead set against computers," Fred Wirth recalled with a laugh. "I mean, he just hated computers. He'd say, we could just hire a bunch of monks; they could do much better." Wirth was one of the Pioneer veterans, joining the project in 1964 to handle data processing tasks for *Pioneers* 6 through 9. Along with Richard Fimmel, Wirth modernized the processing of the science data that began pouring in after *Pioneer* 6's launch. "They made analog wide-band tape recordings, shipped daily, big 9,600 foot, 17-inch reels. We received vast shipments from overseas via JPL, truckloads of tapes. I designed the tape processing station that would take those analog tapes, play them back, and digitize them and place the digital data into a computer. From there we processed the digital tapes for science and engineering data on an IBM 7094."

Still, the process left a lot to be desired. "Between '64 and '68 we were also getting *Pioneer* 6 through 9 data from JPL via teletype," Wirth recalled. "Our first control center was in the lobby of a building at Ames. It was screened off with great big curtains and behind the curtains were 20 teletypes rattling away, spitting out octal numbers of the Pioneer data. We had a whole bunch of guys who plotted the octal numbers with little conversion slide rules and converted the octal numbers to data. I designed a teletype interface for an SDS-10 computer and from then we could get real quick printouts of all this stuff."

Teletypes were still noisy, awkward, and power hungry, however. "Finally in 1969 I convinced Charlie that hey, we oughta get away from this whole teletype business, and I got a piece of equipment through JPL and set up a high-speed data interface directly to JPL. Then of course we could get high-speed data with a real computer to display things and print out. Later on when *Pioneer 10* was built, right at launch, we got Xerox Sigma 5 computers from TRW that they used for testing and building the spacecraft. I got a contractor to program the Sigma 5 computers for project engineering and science data, display and control command system, and display data on the screen and also generate engineering data tapes that we mailed daily to the PIs." Wirth was also responsible for designing and running the Pioneer Mission Operations Control Center at Ames, and by the time

Pioneer 10 reached Jupiter, the project's computer support was state of the art.

Sometimes, however, the human factor could cause problems. During the *Pioneer 10* Jupiter encounter, Wirth said, "my biggest thing was keeping the data flowing, keeping the computers operating. I made sure the data flow was uninterrupted." One day, shortly before encounter, Wirth was sitting at his console and watched with horror as his screen suddenly went completely blank. "We had a glass window near my console in the mission control center to another office where we put up a whole bunch of terminals and the VIPs could sit there and look at data on terminals and look through the glass window into the mission control center." *Pioneer 10* attracted quite a few luminaries, among them Wernher von Braun, who happened to be present in the observation area. "I look through the glass window," Wirth remembered, "and I see Von Braun sitting on the desk, and his butt pushed the button on the video terminal and turned it off. And with that system we had at that time, if one terminal goes down, it crashed the computer. So I ran in there and said, 'Excuse me sir. Would you please not do that? Be careful,' and I turned it back on again, and we had to reload and reboot the computer."

Another aspect of the Pioneer computer setup that helped to both fuel and mitigate the Ames-JPL rivalry is that, from the beginning of the program, the project still had to rely on JPL's computers to actually generate, format, and transmit to the DSN the commands for controlling the Pioneer spacecraft. Although the processing of data from the science instruments and the control of the missions were in Ames hands, the mission controllers at Ames had to tell JPL whatever the spacecraft had to do—fire a thruster, turn on an instrument, whatever—and JPL technicians had to use their own computers to get it done. It was both a remnant of the days when JPL had enjoyed uncontested authority over all deep-space missions and an inescapable necessity because JPL had the only computers and software that could do the job.

The setup required extraordinary coordination among all those involved. Elaborate lists of command sequences and timings, operational plans, and precise timelines were distributed by Ames to JPL and the DSN stations and checked and rechecked constantly to keep everyone on the same page and to ensure that the proper commands were sent to the spacecraft at the proper times and in the proper order. Intensive training sessions, readiness meetings, and simulations were held for months before planetary encounters.

But not long after the die was cast on *Pioneer 11*'s final Jupiter/Saturn trajectory, Hall took a giant step in gaining more independence from the project's technical reliance on JPL. In May 1974, Ames successfully tested a system called the "Pioneer Direct Mode Command System," using two PDP-11 computers in the Pioneer Mission Operations Control Center. As the Ames internal newsletter explained, the most important difference between the old JPL-reliant era and Pioneer's new independence was that "the ability to transmit a command is no longer dependent on the availability or status of the computer systems at the Jet Propulsion Laboratory," also noting that the system would become fully operational by the following January.

The newsletter didn't mention whether there had been any gloating in the hallways of Ames or grumbling at JPL. But no doubt Charlie Hall was very happy.

THE ENCORE PERFORMANCE

Pioneer 11 began humankind's second Jovian encounter in early November 1974, when it entered the outer boundary of Jupiter space. Three weeks later it hit the bow shock and crossed into Jupiter's magnetosphere. A day or so later *Pioneer 11* actually crossed the bow shock again and found itself briefly outside the Jovian magnetosphere even as it steadily closed the distance to the planet: the pulsating magnetic fields of Jupiter had been pushed inward toward the planet by the solar wind. Then *Pioneer 11* went through the bow shock again and reentered the magnetosphere. The same thing had happened to *Pioneer 10*, and *Pioneer 11*'s experience proved that it hadn't been an anomaly.

Jupiter loomed ever closer, receiving its second visitor from Earth in a year. This visit, however, would be considerably different from the first. *Pioneer 11* was going to be a bit wilder, a bit more daring, than its older sister, and it was going to look at Jupiter in ways that *Pioneer 10* had missed. Not only was it going to pass much closer to the planet, it would do so at a much faster speed and against the rotation of Jupiter, allowing it to observe a complete rotation of the magnetic field and radiation belts. And *Pioneer 11* would take a look at the polar regions of Jupiter, something precluded by *Pioneer 10*'s trajectory. If the atmosphere at the poles was less turbulent and more transparent than in the more equatorial regions, it might be possible to see a little deeper into the interior than with a more straight-on approach. *Pioneer 11*'s encounter with Jupiter might be something of a repeat performance, but it would be anything but routine and certainly not boring.

Pioneer 11 crossed the orbits of Jupiter's outer satellites, the bow shock still fluctuating in and out past the spacecraft several more times. Back home at Ames the intense atmosphere of a year earlier returned as the PI teams and the press descended once more on the center. It was *deja vu* all over again, yet with some important differences. If there wasn't quite the same feeling of excitement now, the knowledge of doing something extraordinary for the very first time, there was the sense of confidence that comes from past experience. Hall, the scientists, the engineers, the controllers had already succeeded once, more spectacularly than they may have dared to hope. Now they were going to do it all over again, but this time taking a few more risks and trying to fill in the gaps left by *Pioneer 10*'s earlier triumph.

Although identical to *Pioneer 10* at first glance, *Pioneer 11* sported a few minor differences. There was an extra instrument called a fluxgate magnetometer, added to supplement and enhance the magnetic field measurements of the helium-vector magnetometer carried by both spacecraft. Scientists also managed to incorporate some minor modifications to their instruments devised after the experiment designs had been frozen for the *Pioneer 10* mission. Van Allen, for one, managed the tricky task of modifying his Geiger tube telescope without changing its size, weight, or power requirements. "After the successful launch of *Pioneer 10*, I had some different ideas that I wanted to incorporate in my instrument which had been actually disapproved by the original selection panel," he said. "But I made a special case to reintroduce those, substitute them for another element of my instrument, to accomplish what I wanted to accomplish originally. And that was approved the second time around, so I had an improved instrument on *Pioneer 11*. I had to shoehorn it in the instrument package without changing the mass or power requirements or telemetry requirements. It was quite a job to figure out how to do that." *Pioneer 11* also received a different set of RTGs, which according to some, weren't quite as good as those installed on *Pioneer 10*. Although the power output of the RTGs was known to decrease over time as a consequence of the heat and radiation the devices generated, *Pioneer 11*'s RTGs had displayed a somewhat greater drop in power over the first year of its journey. It wasn't enough to threaten the mission, and the degradation of output seemed to taper off in the months before the Jupiter encounter, but it was still clear that one RTG was not necessarily as good as another: two identical RTG units might not last for the same length of time.

The hard-won experience of *Pioneer 10* in Jupiter's bath of radiation paid off as *Pioneer 11* approached the planet. Now that they knew for certain that the radiation environment would likely interfere with the operation of the spacecraft and its instruments, controllers fed *Pioneer 11* a steady stream of commands designed to compensate for false commands caused by the bombardment of charged particles. *Pioneer 10* had failed to capture some images of Jupiter and its moon Io because of spurious radiation-generated commands to the imaging photo polarimeter, but that wasn't going to happen this time—at least so the scientists and controllers hoped.

But if Hall and the Pioneer team could command and control their spacecraft 500 million miles from Earth, they had considerably less control over events closer to home. The three 64-meter dishes of the DSN were spaced around the globe in roughly 120-degree intervals so that together they would cover the entire sky. No matter where *Pioneer* or any other spacecraft might be in space, at least one of the big DSN dishes would always have it in its sights. But if one of the DSN stations was down for whatever reason and the target spacecraft happened to be in the part of the sky covered by that station, communications would be interrupted or even completely lost. It didn't happen often, and even when it did, it was usually possible to switch communications to another smaller antenna so that contact could be maintained, if at a slower data rate. At least that was true of most spacecraft. To hear the faint whispers of the weak signal from a spacecraft as far away as *Pioneer 11*, the big DSN antennas were needed.

Usually when a DSN station had problems, they were technical in nature: data processing problems, broken communications lines, frozen antenna bearings. But as the time of *Pioneer 11*'s flyby of Jupiter neared, the Canberra station was faced with something far more human and therefore much harder to correct: a possible strike by diesel generator operators that threatened to close down the entire station. And if Canberra was out of commission, Earth would effectively be out of touch with *Pioneer 11* during the most critical phase of its mission thus far—periapsis with Jupiter.

The operations crew scrambled to save the mission. While naturally there was nothing they could do to resolve labor problems on the other side of the world, they could make special technical arrangements so that the Goldstone DSN station would be able to stay in touch with *Pioneer 11* longer than normal, even if it meant reducing the rate of data transmission because Jupiter and *Pioneer 11* would be setting in the California sky at that time.

Fortunately, the situation at Canberra eased and cooler heads prevailed. The DSN would still have full coverage of *Pioneer* during its entire Jupiter encounter, and everyone at Ames breathed a collective sigh of relief. According to Fred Wirth, however, the incident, while unusual in its severity, wasn't exactly a surprise. "The tracking station in Australia is nationally owned," he explained. "They would get the money from the DSN, but it was a national thing with their own employees. It was a running joke that every single launch or encounter or whatever, the Australian tracking station would go on strike or threaten to strike because they knew they had us by the balls. You've got a launch or encounter coming, and you can't delay it, and they were gonna strike."

With this particular threat past, everyone at Ames could get back to the normal chaotic and yet highly focused phenomenon of a planetary encounter. It was easier for some than others. "We tried to forget about everything else and focus just on the job," Richard Fimmel said. "You have questions to answer. You have news media bothering you for information. Problems to solve. Tracking stations had prob-

Principal investigators John Simpson and James Van Allen (sporting a new beard since *Pioneer 10*) compare notes on *Pioneer 11*.

lems. It was a very hectic time, but a very gratifying and exciting time. You'd get a thrill out of what you were able to accomplish because it had never been done before."

Pioneer 11 would make its closest pass to Jupiter, 26,725 miles above the atmosphere, at just after 9 P.M. Pacific Time on December 2. But in another different wrinkle from the *Pioneer 10* encounter, *Pioneer 11* would be out of contact with Earth during its periapsis—it would occur while the spacecraft was passing behind Jupiter. Again the anxious questions: Would *Pioneer 11* still be alive when it emerged from behind Jupiter's shadow, or would the plunge through the radiation belts, so much closer to the planet than *Pioneer 10's* trajectory, be its death warrant?

And again, the waiting. Beginning at 9:42 P.M., the signals from *Pioneer 11* abruptly ceased, having been transmitted about 40 minutes earlier when the craft went behind Jupiter. Because of the light-speed time delay, Earth would not hear from the spacecraft again until about 10:24—if it had survived, of course. More slurping of coffee, chain smoking of cigarettes, tapping of fingers, placing of nervous and optimistic bets.

Ironically enough, the Canberra DSN station that just hours before had caused such anxiety over the fate of the encounter now heralded its success. Its antenna was the first to pick up *Pioneer 11* as the craft sent news of its survival back home, to the cheers of everyone at Ames. Some of the instruments had malfunctioned briefly during the onslaught of radiation at periapsis as with *Pioneer 10*, but they were now returning to normal as radiation levels fell off farther from the planet. Revisiting his "dragon" metaphor of a year earlier, NASA official Robert Kraemer told the press that *Pioneer 11* "flew into the fiery jaws of the dragon and got scorched a little, but it is a tough little bird and it is headed for Saturn."

The gamble had paid off. The different trajectory, polar instead of equatorial in its orientation, had limited the radiation exposure of the spacecraft while allowing it to pass much closer to Jupiter than *Pioneer 10* had and still survive. The boost from Jupiter's momentum and orbital velocity sent *Pioneer 11* upward out of the ecliptic plane on a long arc to its next destination.

As much of a long shot, literally and figuratively, as sending *Pioneer 11* on to Saturn might be, the success of the Pioneer Jupiter missions made the decision seem perfectly logical. *Pioneers 10* and *11* had already proven themselves capable of doing not just the expected and the planned but much more. There seemed to be almost no limit to what these spacecraft could do under the leadership of Charlie Hall and his team. Adding one more planet to the flight plan, even one that had never been visited before, hardly seemed unreasonable.

THE NEW JUPITER

At one time, one scientist remarked, everything that was known about Jupiter would barely fill one chapter of an astronomy textbook. After *Pioneers 10* and *11* visited the planet, there was suddenly enough to fill an entire shelf of books. And as the PIs crunched their data, even more details emerged, dispelling the fog of centuries of ignorance and mystery.

The basics, of course, were well known to every high school science student. Jupiter is the fifth planet from the Sun and the largest in the solar system, orbits the Sun in a period of about 12 years, spins rapidly with a day only about 10 hours long, features bands of colored clouds and a Great Red Spot, and was first observed through a telescope by Galileo Galilei, who also described the planet's four largest moons—Ganymede, Europa, Io, and Callisto. It is also a strong source of radio signals, indicating the presence of charged particles trapped inside a magnetic field. The planet is composed mostly of hydrogen and helium, leading some astronomers to argue that it had more in common with a failed star than a finished planet.

Before Pioneer that was about all that was known about Jupiter, at least in the broad strokes. The rest was all enigma. The inner structure of the planet, the workings of its atmosphere, the nature of its moons, the mechanisms by which it seemed to emit more energy than it received from the Sun were all open questions. Although quite a lot could be deduced from telescopic, spectroscopic, and radiotelescope observations from Earth, the only way to truly answer all the questions was to actually visit the planet and make direct observations from close up, instead of from over 360 million miles away, which is as close as Jupiter and Earth ever come together.

In an article he wrote before the Pioneer encounters, the University of Chicago's John Simpson pointed out that Jupiter seemed to be "a very primitive planet" and that it might turn out to be "a new Rosetta Stone, enabling us to go back in time to reconstruct much of the chemistry and physics of planetary formation. . . . These studies may cast light on the origin of our solar system."

The Pioneer missions showed Jupiter to be much more beautiful, dangerous, and complex than anyone had imagined. The IPP images revealed a vastly intricate, practically liquid planet with no discernible solid core, composed mainly of hydrogen and helium and various other elements that revealed themselves in the colors of Jupiter's clouds. The planet proved both larger and more massive than expected, with a restless atmosphere of clouds of ammonia, methane, traces of water vapor, and other substances, churning about Jupiter at high velocities and in complex convective patterns. Pioneer showed that the famous Great Red Spot is an atmospheric phenomenon, a huge hurricane-like storm that had persisted for centuries because, unlike on Earth, Jupiter has no land masses to break up atmospheric disturbances. And while on Earth the weather is mainly driven by solar energy, so that weather patterns change because of temperature differences at different latitudes and the light and dark sides of the planet, on Jupiter the Sun's impact on the weather is negligible; atmospheric activity appears to arise due to heat generated by processes deep inside the planet.

In fact, Pioneer showed that Jupiter gives off about twice as much heat as it absorbs from the Sun. While the temperatures at the cloud tops is about −234 degrees Fahrenheit, the center of the planet might be as hot as 54,000 degrees. Apparently that heat was left over from when the planet formed. *Pioneer 11*'s polar flyby observed that the cloud tops are lower at the poles than at the equator—the liquid planet bulges outward in its midsection.

This internal heat and the turbulent broiling of hydrogen and other elements, possibly in a strange metallic form deep inside the planet, create immense electrical currents in the core that induce Jupiter's magnetic field and its huge and complex magnetosphere. Pioneer demonstrated the Jovian magnetosphere to be much more intense and extensive than previous models had predicted, stretching far into interplanetary space. Its weaker outer regions are subject to distortion by the solar wind, as the Pioneers discovered when they encountered repeated crossings of the bow shock as the outer magnetosphere was pushed inward and then expanded outward again. The inner radiation belts of trapped charged particles that had caused such anxiety over spacecraft survival had also proved much more intense and extensive than expected. Yet the inner moons of Jupiter, particularly the Galilean satellites, act to moderate the radiation flux, absorbing particles as they orbit around the planet. Radiation levels in the wake of a Jovian moon are much less intense than elsewhere near the planet. Such particle flux absorption effects observed by *Pioneer 11*'s flux gate magnetometer also provided the first tentative hints of Jupiter's faint ring system.

Although their primary attention was directed toward Jupiter, the Pioneers managed to make some tantalizing observations of some of the Jovian moons. *Pioneer 10*'s occultation of Io, when it passed briefly behind the satellite, allowed radio signals to be affected by the moon and showed that Io has an atmosphere and is embedded in a cloud of hydrogen. It also has an ionosphere, odd for such a small world. Obviously there was a lot going on at this strange little moon, but the details would have to wait until Voyager's later, more extensive observations. Pioneer also provided preliminary data on the density and composition of the Galilean moons, not to mention some fairly crude but still intriguing images, all of it information that would be of immense value to the team planning the Voyager encounters.

One of the most startling discoveries of Pioneer, especially to the particles and fields specialists, was that Jupiter is a source of high-energy bursts of electrons that it sprays throughout the solar system. These bursts had been detected both at Earth and by spacecraft in the inner solar system, even as far from Jupiter as the orbit of Mercury, and had been a nagging mystery for years. Scientists realized that they weren't coming from the Sun as might be expected, so what was causing them? Pioneer settled the question at last.

And William Kinard's "air mattresses" found that there was indeed more interplanetary dust in Jupiter space than in open space—about 300 times more, in fact. Fortunately, the particles detected were much too tiny to prove much if any threat to a spacecraft. The particles that were worrisome to visitors to the realm of Jupiter were more of the charged atomic than the microscopic dust variety.

As *Pioneer 11* left Jupiter far behind and began the lonely trip to its next and final destination, NASA administrator James Fletcher officially changed its name from *Pioneer 11* to *Pioneer Saturn*. It was perhaps a bit optimistic. Although the spacecraft was now inexorably bound for Saturn, it was quite possible that it wouldn't have much to say when it arrived because there was a good chance it might not be operating by then. On the other hand, Pioneer inspired optimism.

A Jewel in the Night

With *Pioneer 11*'s departure from Jupiter, life quieted down to a routine at mission control. *Pioneer 10* had settled comfortably in its extended mission, continuing to probe the interplanetary medium, solar wind, and galactic cosmic rays. Controllers at Ames kept close tabs on the spacecraft's well-being, of course, but there was rarely if ever any excitement to report or emergencies to contend with, just the steady stream of data that was duly processed, transferred to tapes, and passed on to the various principal investigators.

A slight mystery arose with *Pioneer 11* a few weeks past Jupiter. Somehow, erroneous commands were being generated to some of the instruments and were interfering with their proper operation, much as had happened with both Pioneers during their Jupiter encounters. Then the cause had been the intense radiation bombarding the spacecraft's electronics. Now, however, *Pioneer 11* had left the Jovian radiation hell far behind, surrounded only by the normal background of deep space. Obviously, something else was causing the problem. Diagnostic tests of all the instruments, including turning each one on and off in turn, finally demonstrated that the Sisyphus asteroid-meteoroid detector, again damaged by Jupiter's radiation as had been its twin on *Pioneer 10*, was the culprit. Once the Sisyphus instrument was turned off, the spurious command signals ceased and the other instruments returned to normal functioning—all but one, *Pioneer 11*'s plasma analyzer, which for some reason continued to malfunction after being reactivated. The instrument's loss would be a serious blow to the Saturn encounter, but with Saturn still almost 6 years away, at least there was still plenty of time to correct the glitch.

On the way, *Pioneer 11* would keep busy by studying the solar wind and interplanetary medium like its sister craft, although from its much different vantage point of outside the ecliptic plane. Much as the different perspectives of *Pioneers 6* through *9* provided a fuller and more complete picture of the space environment of the inner solar system, *Pioneers 10* and *11* would widen the canvas to include its outer reaches.

For the next several years, however, activity around the Pioneer offices and control centers would be fairly quiet. Plans were coming together for what would prove to be the last mission, Pioneer Venus, to be launched in 1978. It was an ambitious multiple-spacecraft mission, and preparations for it began to command increasingly more of the attention and labors of Charlie Hall and other Pioneer personnel at Ames, even as *Pioneer 10* continued out of the solar system, crossing the orbit of Uranus in July, and *Pioneer 11* proceeded toward its Saturn rendezvous.

But none in the Pioneer community, not Hall, his team, and especially not the PIs, ever let the impending Saturn encounter stray too far from their thoughts. Indeed, years before *Pioneer 11* arrived at Saturn in September 1979, the mission became the focus of a scientific and administrative controversy that had profound implications not only for the survival and success of *Pioneer 11*'s Saturn encounter but also for the fate of the upcoming Voyager mission.

THREADING THE NEEDLE

Although they might not like to admit it openly, the Pioneer team always realized that in many quarters of NASA and elsewhere, their missions were considered to be little more than a humble prologue to the main show, and that *Pioneers 10* and *11* were just crude pathfinders, understudies for the upcoming true stars of outer solar system exploration: JPL's *Voyagers 1* and *2*. To some of the Pioneer people, it seemed that the Voyager team had decided that if their spacecraft had to be beaten to Jupiter and Saturn by the low-rent Pioneer, then the Voyager acolytes would do their best to ensure that Pioneer did everything possible to reduce Voyager's risks and enhance its chances for greater success.

Officially, of course, at least as far as NASA headquarters was concerned, it was definitely true that Pioneer was always intended to serve as a pathfinder for future missions. In a pre-Jupiter encounter *Pioneer 10* press conference in February 1973, NASA's associate administrator for space science, John Naugle, told reporters that both *Pioneers 10* and *11* were regarded as "precursor missions." Their task was simply to fly by Jupiter and "lay the groundwork for the outer planets exploration program which we see beginning with the Mariner-Jupiter-Saturn mission [Voyager] in 1977." Instead of the alpha and the omega of NASA's deep-space effort, the Pioneer project was only a small and preliminary part of a much grander program of solar system exploration.

If such pronouncements by NASA hierarchy weren't enough to give the Pioneer team something of an inferiority complex, there were the efforts of some to influence Pioneer mission planning before the spacecraft ever left the ground. JPL scientists planning for the future Voyager missions had proposed a trajectory for *Pioneer 10* that would send it within 1.3 Jupiter radii (R_j), which was the close approach needed to obtain the gravity boost to send a spacecraft on to Saturn. The reasoning was that if *Pioneer 10* survived intact, Voyager's scaled-down Grand Tour would be proved feasible. But Pioneer scientists insisted that such a course would be too risky, because Jupiter's radiation environment might be intense enough to destroy *Pioneer 10*'s electronic systems and render the craft inoperable. Instead, they argued for a flyby at about 3 R_j to increase the spacecraft's chances for survival. Besides, *Pioneer 11* would follow its sister craft a year later and could always be sent on a closer approach if *Pioneer 10* emerged intact. The Pioneer team won this particular argument, and *Pioneer 11* later made the closer pass that amply demonstrated the feasibility of Jupiter gravity assist. Thanks to NASA's policy at the time of using two spacecraft for planetary missions, the needs of both the Pioneer and the Voyager science teams could be met—at Jupiter.

Unfortunately, only one Pioneer spacecraft was going to Saturn—unfortunately, because again there was controversy over the best trajectory. Saturn had a set of huge rings, gorgeous but also a threat to spacecraft. Upon arrival at Saturn, *Pioneer 11* would face perhaps the most dangerous moment of its entire voyage, the crossing of the planet's ring plane. There were two choices: *Pioneer 11* could plunge into the unknown zone between Saturn and its visible rings or steer a more prudent course outside of the rings.

This time it wouldn't be possible to have it both ways. A ring plane crossing had to be made, either outside or inside the rings. Ironically, the respective positions of the Pioneer and Voyager projects during *Pioneer 10*'s Jupiter encounter would be reversed for *Pioneer 11* at Saturn. The Pioneer scientists were willing to risk a close approach within the rings, because even though it might prove fatal to the spacecraft, it held the promise of obtaining unique data. But the JPL team, now planning to extend *Voyager 2*'s mission to Uranus, pushed for a *Pioneer 11* outside-the-rings flyby to determine the possible hazards of *Voyager 2*'s passage. If Voyager was to use Saturn to boost itself to Uranus, it had to pass by the ringed world at a specific point, which happened to be outside the ring plane. John Simpson stated flatly that the trajectory issues "were very serious, and they represent the kind of interplay and debate that goes on typically within NASA." The question of which choice would yield the greatest scientific return, not to mention the best chance for the survival of both spacecraft, was made even more complex because little was known and much was controversial about Saturn's rings.

Saturn's three main visible rings, comprising an outermost A ring, a middle B ring, and a diaphanous "crepe" C ring closest to the planet, along with the Cassini Division separating the A and B rings, had been known for centuries. Yet most scientists believed the ring system to be much more intricate and to extend beyond the boundaries that could be detected from Earth. In 1969 the French astronomer Pierre Guerin claimed the discovery of a tenuous D ring inside the known rings, separated from the C ring by a small gap called the Guerin or French Division. Some other astronomers supported Guerin's hypothesis, although with somewhat ambiguous data. Outside the A ring, there was also evidence of a possible E ring that might extend into space more than twice the diameter of the visible rings.

But the exact composition of Saturn's rings and most importantly the average size of the ring particles were unknown. If the particles were smaller than about 1 millimeter, they wouldn't be able to penetrate a transiting spacecraft; if they were larger than about 1 centimeter, the particles would likely be spaced widely enough apart that a craft would pass through them without any impacts. However, ring particles in the range of 1 millimeter to 1 centimeter would be massive enough and close together enough that a lethal impact would be almost inevitable, particularly given the great speed of the spacecraft. It was the asteroid belt scenario all over again, but this time the doomsayers had a point. The rings were definitely there, and they filled much less space than the asteroid belt. There was no question that the rings were infinitely more hazardous than the vast asteroid belt.

At a workshop on Saturn's rings held at JPL in 1973, scientists debated both the existence of the D ring and the prospect of using *Pioneer 11* to settle the question for Voyager. Kitt Peak Observatory's Lyle Broadfoot said, "We are going to go as close to the rings as we can. The real question is how close is safe." NASA's John Niehoff suggested that *Pioneer 11* could be committed to a "kamikaze-type passage to the edge of the A ring," the point that Voyager would have to pass in order to go on to Uranus. The risks to any Saturn-bound craft were real, and although data could be compiled, models could be constructed, and odds could be calculated, the only way to definitively settle the question was to send a spacecraft through the ring plane.

The debate over *Pioneer 11*'s Saturn targeting became a public issue as early as May 1976 in a NASA press release which stated plainly that *Pioneer 11* would "pass either through the dark space between the inner ring and the visible surface of the planet—or it will fly outside the rings." Emphasizing that "uncertainty in the final selection of flight path depends almost entirely on unknown characteristics of Saturn's rings," the release made no mention of the Voyager factor, although it seemed to side with the Pioneer team by pointing out that "nearly all experiments would benefit from the closest possible fly-by of the planet."

The course adjustments for the chosen trajectory would have to be made by mid-1978 at the latest, so a decision could not be long delayed. By early 1977 the question had settled around two distinct choices: the "inside option," which would aim for a point between the visible rings and the planet, through the hypothesized D ring or possibly the Guerin Division, and the "outside option," crossing the ring plane outside the A ring at about 2.87 Saturn radii (R_s), the point through which *Voyager 2* would later have to pass in order to receive the gravity assist to proceed to Uranus.

In January 1977, Pioneer project leaders at Ames Research Center prepared a preliminary proposal in support of the inside option. The inside track might provide valuable views of Saturn's possible magnetosphere, radiation belts, and the interaction between the rings and charged particles in the Saturnian system, information that even the more sophisticated Voyager craft would not be able to acquire because of its more distant flyby. And if, as headquarters and the Voyager project maintained, Pioneer was chiefly intended to be a preliminary exploratory survey anyway, there seemed to be little to lose and much to gain by choosing the riskier inside course. As Charlie Hall explained, "Anything that we got after Pioneer at Jupiter was gravy. So we go through the rings of Saturn and get smashed up. Well, that's what you went to find out, how solid they are."

The JPL Voyager team, however, had a decided preference for the outside option and, in an unusual move, appealed to NASA headquarters to argue its case. In a letter to Thomas Young, NASA director of planetary programs, a month after the Pioneer proposal, Voyager project manager John Casani pointed out that in order to optimize the scientific returns for Voyager, the dangers of the possible outer E ring needed to be established. He wrote, "If we knew this crossing point was too hazardous, we would give up Uranus and reoptimize the . . . trajectory conditions at Saturn." In other words, if Pioneer did not survive a ring plane crossing outside the A ring, Voyager's first-choice scenario of continuing on to Uranus would be abandoned, and alternative plans could then be made for a more extensive mission at Saturn. Author Mark Washburn describes the situation in his book *Distant Encounters*: "If Pioneer made it through, then presumably Voyager would as well. If Pioneer crashed, *Voyager 2* could be retargeted. It was rather as if Lewis had told Clark to check out a cave for grizzly bears."

Tom Young assigned David Morrison, assistant deputy director of lunar and planetary programs, to coordinate the trajectory deliberations. After studying the arguments of both camps, Morrison issued a rather equivocal memo, finding evidence to support both trajectory options. In support of the inside option, Morrison said, "The opportunity to explore inside the C ring of Saturn is unlikely to present itself again in the foreseeable future. . . . Thus, the spirit of adventure and exploration leads us toward the inner trajectory—it is our only chance to measure directly the ring inside the C ring, and if by chance the spacecraft should survive, there would be a major bonus in information. . . . The basic rationale for going inside remains one of exploration—*the possibility of finding something unique and unexpected* [emphasis in original]."

But in considering the possible total scientific gains to be had by a Voyager Uranus mission, Morrison seemed to conclude that ensuring Voyager's survival was rather more important. "Certainly, *if* a trade must be made, more science should be obtained from a successful Voyager encounter (at Saturn) than from a successful Pioneer encounter," he wrote. "There appears to be a small but significant preference for the outside option, based both on Pioneer science by itself and on the combined value of the Pioneer and Voyager investigations of Saturn."

In a September 1977 memo to Pioneer project staff and PIs attached to a copy of Morrison's memo, Charlie Hall observed that even the PIs had not reached agreement on the question. Reviewing the opinions he had received to date, he noted: "We do not find agreement for the inside option in these responses; rather they are divided and can be interpreted to slightly favor the outside option. Most responses included recognition and acceptance for the alternative to their own choice; and the few unqualified preferences were divided."

The position of each of the Pioneer scientists at this time was largely a function of the nature of his own experiment. The particles and fields specialists such as John Simpson and James Van Allen supported the inside option because such a trajectory would provide the best evidence to determine if Saturn had a magnetic field, not to mention the chance to observe interactions of any trapped charged particles with the ring system and the magnetosphere. They believed that the chance of *Pioneer's* destruction was outweighed by the possibility of new discoveries— the "spirit of adventure and exploration," as David Morrison had phrased it. The PIs whose experiments were not directly affected by the trajectory choice were largely noncommittal, except for Frank McDonald. "I was strongly against [the inside option]," he said. "Everybody else wanted to go in there. I still don't understand their logic. I wanted to live to fight another day." Others were more equivocal. As one scientist put it, "I still have a slight, undecided preference for the outside option, although I cannot say I would be really unhappy with the alternative." Years later Simpson maintained that among the Pioneer PIs there was even some under-standing, if not outright support, of the JPL/Voyager position: "That was their [JPL's] obligation, to see that what Pioneer learned was use-ful for Voyager."

The press had caught on to the story by this time. An October 1977 *Science News* article echoed Charlie Hall's memo when it re-ported: "A *Science News* poll of the 13 principal [Pioneer] scientists [3 of the 13 were unavailable but had previously expressed opinions] was split right down the middle: 6 for the outside option, 6 for inside, and 1 neutral. And even those views were highly qualified. . . . Con-tributing to the tension is that fact that, until the encounter actually takes place in early September of 1979, no one will know whether the decision was the right one. . . . The day of decision looms."

Hall realized that the Pioneer project would have to settle on a unified final position for presentation to Tom Young and NASA head-quarters and scheduled a meeting of all concerned parties for No-vember 1 at Ames. As he recalled: "Each of the scientists had a thing they wanted to do. The problem was to get all these wants into the same bag and see if you could solve the trajectory."

Two representatives from the Voyager project, Ed Stone and Charles Kohlhase, advanced the argument for the outside option at the meeting. Pioneer's Jack Dyer remembered: "They emphasized that our craft would provide better science by doing what they wanted. They felt that since our spacecraft was a low-cost forerunner, it was appropriate for it to be used to validate Voyager's flyby." There was that vague attitude of dismissal toward Pioneer again. But Dyer re-calls that most of the Pioneer personnel, at least among themselves, felt that even if the inside option was chosen and *Pioneer 11* was de-stroyed without pathfinding for Voyager through the E ring, the Voy-ager team would still take the risk and go for Uranus anyway. Dyer added that some on the Pioneer team believed that JPL and NASA headquarters were "embarrassed that this cheap spacecraft could go out there and do all this." In a cover letter to Thomas Young included with his meeting report, Charlie Hall argued that "the Voyager Project's preference for a pathfinding mission . . . appeared to be the only persistent basis for urging an alternative trajectory. . . . For us, the possibility of extending to Uranus with a Voyager class of spacecraft would be almost impossible to vacate even in the unlikely event of a failure of *Pioneer 11* in the E-ring."

The Pioneer PIs each presented their individual viewpoints on the inside/outside options as the choice related to their own scientific objectives. After general discussion and presentation of the latest data on the D and E rings and the possible hazards of each, the PIs voted on the question. Only one of the 12 PIs preferred the outside option: Frank McDonald. The group consensus was that *Pioneer 11* should be targeted for a flyby inside Saturn's rings through the Guerin Division. They summarized their rationale for the choice in the meeting report: "The predictable science return is essentially the same on half an inside trajectory (assuming *Pioneer 11*'s destruction) as a complete outside trajectory. But the inside trajectory gives particularly the possibility of unique results from the particles and fields instruments which cannot be duplicated by Voyager in any case because Voyager doesn't get that close to Saturn; and it is in a sense more conservative because you do not even run the risk of failure by ring crossing until you have returned those crucial data including the possibilities from important serendipitous discoveries."

The group felt that the scientific promise of the inside option was ultimately of more importance than the survival of the spacecraft. "Preservation of *Pioneer 11* by means of a low-risk flyby of Saturn is of relatively limited value because subsequent science will be accomplished to greater distances by Voyager. . . . Choosing the inside provides some probability of finding something new. . . . Pioneer should obtain particle and field data inside the rings where Voyager cannot."

Essentially, the Pioneer team felt there was nothing to lose by going for broke. The humble but hardy *Pioneer 11* had already accomplished its primary mission at Jupiter and achieved amazing things, much more than had ever been expected of it. Even if the spacecraft didn't make it through the D ring, it would be a fitting tribute for the little ship to go out in a blaze of glory while making one more important discovery.

Hall's summary of the meeting, along with a formal "Request for Approval of Pioneer Targeting at Saturn," went to Thomas Young at NASA headquarters on November 8. Several weeks later Young held a long-distance telephone conference with Pioneer project personnel at Ames and Voyager team representatives at JPL. John Dyer recalled that he and the rest of his Pioneer colleagues were hopeful for a decision in their favor, particularly because in the course of the phone conference, Young's deputy, David Morrison, seemed to have altered his previous leanings toward the outside trajectory in favor of the Pioneer proposal. But after spirited discussion from all quarters, Young finally announced his decision: *Pioneer 11* would be sent outside Saturn's rings. Charlie Hall vividly remembered the reaction at Ames to Young's decision: "I never heard so many boos."

In a subsequent letter to Hall, Young explained that while he was "personally impressed by their arguments" and that he could "appreciate why the investigators so strongly favor the inside choice," he had chosen the outside option because the long-range aims of solar system exploration had to outweigh the more immediate preferences of an individual project. He wrote that "given the commitments we have made to the Voyager Project, and the uncertainties associated with conditions at ring plane crossing at 2.87 R_s, it is essential for us to do everything we reasonably can to ensure Voyager's success. If Pioneer is targeted to 2.87 R_s and fails upon ring plane penetration, we will almost certainly have to reassess our plan to continue to Uranus. . . . Alternatively, a successful Pioneer will greatly increase our willingness to commit *Voyager 2* to the Uranus option. . . . Thus, either survival or non-survival of Pioneer on the outside trajectory can have important influence on Voyager plans, and thus on achieving the maximum science return from all three spacecraft. . . . If there is a significant chance of destruction . . . it is absolutely vital that we obtain all the information we can on this hazard before committing Voyager to the Uranus option."

Several months later in the spring of 1978, controllers at Ames Research Center made the final adjustments to *Pioneer 11*'s trajectory as it closed in on Saturn. The course was set; in September 1979 the actual encounter would prove whether Young's decision had been the best one.

SURVIVAL AND SUCCESS

The year 1979 was a banner year for planetary exploration, perhaps the most remarkable one yet. Since Pioneer's Jupiter encounters, *Voyagers 1* and *2* had been launched in the fall of 1977. In early March 1979, *Voyager 1* reached Jupiter and startled the world with fantastically detailed and beautiful images of the planet and its moons (thanks to its dedicated camera system and three-axis-stabilized spacecraft configuration), along with a wealth of new scientific data that complemented and built on Pioneer's earlier findings. Among many other findings, Voyager confirmed Jupiter's faint ring system and also established what was so odd about Io: it was a hellish little world in constant upheaval from active volcanoes. In July, *Voyager 2* followed up at Jupiter with still more striking pictures and discoveries.

Meanwhile, *Pioneer 11* was heading back into the ecliptic plane, on course for Saturn. Like the rest of the world, the Pioneer team had watched the Voyager encounters with awe and wonder, marveling at the beautiful images of Jupiter that Pioneer had not been equipped to capture. No one could deny that Voyager was a spectacular mission, but neither could you deny that Pioneer had been the first one through, even if the jubilant JPLers sometimes had to be gently reminded of that fact—or sometimes even not so gently. Some of the Pioneer scientists argued that however wonderful the images, Voyager didn't really discover much that was new, and some of its supposedly new findings had already been found upon further analysis of the previous Pioneer data (including, for example, Jupiter's tenuous ring system).

But there was no time for pointless internecine rivalries now. All eyes, whether at Ames or JPL, turned to the impending *Pioneer 11* encounter. The annoying glitch in *Pioneer 11*'s plasma analyzer had been fixed, and the spacecraft and its instruments were ready for the plunge through the ring plane. Before that happened, however, there was the question of Saturn's bow shock and magnetosphere. The Pioneer PIs were fairly certain that Saturn, like Jupiter, had some kind of magnetic field, but its exact nature was more of an unknown than Jupiter's. "The intensity of conjecture grew as the spacecraft approached Saturn," James Van Allen remembered. He made a bet with project scientist John Wolfe on when *Pioneer 11* would cross into the bow shock and Saturn's magnetosphere would be found. It happened on August 31, and Van Allen won the bet. "Being a man of honor, Wolfe immediately wrote me a personal check for 'zero and 50/100 dollars,' the agreed amount of the bet, which represented the confidence in our conjectures as well as the magnitude of the financial risk that an academician was willing to accept," Van Allen said.

In fact, most everything about Saturn was more mysterious. "Anything about Saturn was a fresh discovery," Van Allen said. "In some sense the *Pioneer 11* Saturn encounter was even more exciting. The Jupiter encounter confirmed previous evidence with enormously greater detail, but it wasn't all discovery like Saturn."

The IPP began taking images of the planet in early August. For rookie Pioneer controller Ric Campo, it was a heady initiation. "I sent the commands to return the first Saturn image taken from a spacecraft," he remembered. "About the time that I sent the command sequence, we needed to get in touch with the mission director, who was off-site that day. We paged him, but we did not get a response. The commands came back and we received the image. The mission director came in around that time, and we thought it was funny that we got commands to and from Saturn, yet his page wasn't received."

Approaching Saturn on August 26, 1979, *Pioneer 11* captured this image from a distance of 1,768,422 miles.

As with Jupiter, the first images of Saturn were somewhat disappointing and lacking in detail, with the planet only about the size of a nickel and the rings a faint smudge on either side. That changed quickly as the spacecraft got closer. Soon *Pioneer 11* was able to discern more details of the rings and began making ultraviolet measurements of the atmosphere. Still, Saturn itself was something of an anticlimax after Jupiter's intricately banded and colorful face. Its atmosphere was less variegated and more homogeneous, apparently less storm wracked than Jupiter's cloud bands.

That didn't make the encounter any less exciting, however. Early on the morning of September 1, Gehrels's IPP revealed a new ring outside the A ring, which was dubbed the F ring. On the same image was a new moon, provisionally named 1979 S1 (for the first satellite of Saturn discovered in 1979).

Later that morning came the acid test. Once more, Hall, his con-

trollers, and scientists and engineers waited, huddled over the computer terminals of the cramped Pioneer Mission Operations Control
Center at Ames. *Pioneer 11* was due to cross the ring plane of Saturn
outside of the A ring at 7:36 A.M. Pacific Daylight Time. Traveling at
about 70,000 miles per hour, the spacecraft would pass through the
danger zone in less than a second—unless a collision with ring particles happened first. But as usual, because of the light-speed communications lag, 86 minutes would pass before Pioneer's signals
reached the dishes of the Deep Space Network. If the stream of data
from *Pioneer 11* fell silent, the anxious humans waiting inside the
cramped room at Ames would know that the worst had happened:
Pioneer 11 was dead, and *Voyager 2* would not be heading for Uranus.

The anxious minutes passed, and at 9:02 A.M. the computer
screens again began to flicker with fresh data. *Pioneer 11* was still
transmitting. It had survived. David Morrison described the scene:
"There were scattered cheers and many sighs of relief," although "a
few rueful comments were also made about not having tried for the
target point inside the rings." But the Voyager scientists, some of
whom were also present in the Pioneer control room, were happy.
Their path to Uranus and ultimately Neptune was now open, courtesy of *Pioneer 11*.

Pioneer 11 then zipped just underneath the rings and closed to
within 13,000 miles of Saturn. And then the inevitable and agonizing
communications blackout began as the craft passed behind the planet.
Another 86 minutes of waiting began. At least this time the fear wasn't
that the spacecraft would emerge from the blackout fried by radiation; it had already shown that Saturn's rings very efficiently absorbed
most of the charged particles trapped within the magnetosphere. In
fact, in the shadow of the rings, particle counts dropped off to nearly
zero, making some of the particles and fields experimenters wonder
briefly if their instruments were still working.

The danger was still that of an impact with ring material, either
in the known rings or in another yet-undiscovered new ring. But *Pioneer 11* reappeared on the other side of Saturn alive and well, again

crossing the ring plane without incident as Saturn's gravity looped it upward. It continued on to pass by the second largest satellite in the solar system (second to Jupiter's Ganymede), Titan, on September 2, taking a few pictures and making some brief observations of the moon on the way out of the Saturnian system.

Unknown to anyone back on Earth, the spacecraft had indeed narrowly evaded disaster. Apparently, the very moon *Pioneer 11* had discovered just a day earlier, orbiting close to the planet in a period of only 17 hours, had swung back around to be in the right place to destroy its discoverer—almost. But the narrow escape had also provided yet another Pioneer first: the detection of a moon not by visual data or with a telescope but by particles and fields instruments. It was during the initial analysis of their data sometime later that the PIs realized something had happened. All the particle readings dropped out at the same time for a brief period. Van Allen recalled: "At the next morning's group meeting, I reported this finding and remarked that it was the result of either a diabolically ingenious instrumental failure or the absorption signature of a close encounter with a heck of a big rock."

Once more the phenomenon that Charlie Hall dubbed "Pioneer luck" had won out. It wouldn't be the last time. Or as John Wolfe put it in a press conference when asked for a "ballpark guess" about the probability of *Pioneer 11*'s survival through the ring plane: "Well, if it's like everything else in this program, we'll survive it, of course."

As *Pioneer 11* left Saturn and headed toward interstellar space at over 22,000 miles per hour, ending another successful Pioneer encounter, James Van Allen found that aside from the flood of data waiting for his analysis, he had one more piece of unfinished business. John Wolfe sought him out and complained that he hadn't been able to balance his checkbook. What happened to that 50-cent check he'd written to Van Allen? "Cheer up, John," Van Allen told him. "I didn't cash your check. I framed it and hung it in my office as a memento." Again, all the Pioneer books were balanced.

THE NEW SATURN

Astronomers had long known that the planets of the solar system were of two broad types. First are the small, dense, rocky worlds such as Earth and Mars, and second are the gas giants, of which Jupiter was the prototype. Of the gas giants, which by virtue of their distance were more enigmatic, much more had been known about Jupiter than about Saturn, Uranus, or Neptune, simply because Jupiter was much closer to Earth. To a cursory glance, the gas giants seemed to be fairly similar, with the exception of Saturn with its rings. Certainly there was little reason to suspect that the solar system's lesser giants were more than not so magnificent versions of Jupiter.

Pioneer 11 proved otherwise. Saturn was much more than just Jupiter with some pretty rings. "Everything we found out at Saturn was totally new," James Van Allen said. Of those new discoveries, the most important was the confirmation that Saturn did indeed possess a magnetosphere. While that wasn't so unusual in itself—after all, Earth and Jupiter also have strong magnetic fields—Saturn's version was markedly different in some important ways. For one, *Pioneer 11* found that the axis of Saturn's magnetic fields is aligned with the planet's rotational axis, with the same geographic and magnetic poles. This was more than just a curiosity. Until Saturn, a crucial aspect of the models explaining how planets generated magnetic fields was the polar offset or tilt of the field, thought to be a necessary characteristic of the internal processes that created planetary magnetism. *Pioneer 11* forced planetary geophysicists to do some serious revising of their standard models. Also, Saturn's magnetosphere is not as unsettled as that of Jupiter, and although it also traps charged particles within itself, the rings and inner satellites absorbs much of the particle flux, keeping the radiation environment around Saturn much quieter and less intense than at Jupiter.

Then, of course, there were the rings. *Pioneer 11* sent back the first close-up views of Saturn's rings, and from perspectives impossible from Earth-based telescopes: backlit by the Sun, illuminated from above, caught in shadow. These new views gave vital clues about the composition, thickness, and distribution of material inside the rings. *Pioneer 11* found two new divisions within the ring system that, unlike the Cassini Division, are invisible from Earth, confirming the existence of the hypothesized Guerin or French Division, and showed that these divisions are not as empty as they might appear but rather are filled with ring particles. Despite later claims to the contrary, *Pioneer* saw the F ring with Tom Gehrels's IPP. "People should not do shabby tricks, for instance, claiming that the F ring was discovered by Voyager," he said. "These old eyes right here, and the crew on duty at the time, saw the F ring first. That was done by Pioneer." As was the case with Jupiter, the extremely detailed and more aesthetic images would have to wait for Voyager to arrive with its pair of three-axis stabilized cameras and its other imaging instruments, but Pioneer told them where to look.

Like its Jovian companion, Saturn is composed mainly of hydrogen and helium and radiates more heat than it takes in from the Sun, apparently generated by the same internal pressures and convective processes as inside of Jupiter. Yet Saturn's atmosphere appeared considerably less violent than Jupiter's, without any persistent features such as the Great Red Spot. In general, Saturn seemed to be another "failed star," an essentially liquid planet without any solid surface. Detailed analysis of Saturn's influence on Pioneer's trajectory yielded the most accurate measurements yet of Saturn's shape and gravity field, characteristics that, like similar measurements at Jupiter, would be critical for the computation of future spacecraft trajectories that would use the planets for the gravity assist technique.

Pioneer 11 also managed to take a tantalizing glimpse of Titan, providing some information on the moon's mass and atmosphere, if not any decent images. The prospect of better data on Titan was a casualty of communications problems, which had also interfered somewhat with the entire Saturn encounter, caused by massive solar activity that generated radio noise and forced Pioneer controllers to reduce the data rate between Earth and the spacecraft sometimes to as low as 512 bits per second, from a top rate of 1,024. The situation was made worse by an unfortunate but unavoidable planetary alignment at encounter time. Earth and Saturn were approaching superior conjunction, so that the Sun was almost directly between them. *Pioneer 11*'s already faint radio signals to Earth had to pass through the strong cloud of radio static created by the Sun. If *Pioneer* had reached Saturn only a little bit later, the Sun would have blocked communications completely. But thanks to improvements in Deep Space Network technology since the earlier planetary encounters, as well as a superb effort by DSN personnel in using and adapting that technology to pick out *Pioneer 11*'s weak murmur amid the solar noise and keep their antennas locked on, the link to Earth was maintained.

THE OPEN FRONTIER

With the end of the Pioneer Saturn encounter, the solar system exploration baton was passed finally to Voyager and the other missions that would follow. The pathfinding and trailblazing were over; now everyone would follow in Pioneer's wake. To Charlie Hall, the Pioneer PIs, and everyone who had worked on the project at Ames, TRW, JPL, the DSN, and elsewhere, it felt like the end of an era. It had been quite a journey from Al Eggers's first modest idea of a few small solar probes to Saturn and beyond into interstellar space.

The gainsayers who had scoffed at the prospect of Ames running a space project had been duly and quite convincingly silenced. Hall had also proven that the idea of individual fiefdoms within NASA was nonsense; centers other than JPL were perfectly capable of conceiving, planning, and executing planetary missions. Each individual NASA center might still have its particular areas of expertise and specialization, but those responsibilities didn't have to be a straitjacket. If good ideas for a specific project came from a part of the agency not usually known for handling such a job, they wouldn't be dismissed out of hand simply because the task might be outside that center's usual province. Hall and the Pioneer project hadn't been alone in challenging the conventional wisdom but certainly provided a shining example to others.

Whatever doors Hall had opened for those operating within the internal politics of NASA, however, they were insignificant compared to the grander scientific and technological barriers that *Pioneers 10* and *11* demolished. Defeating the Great Galactic Ghoul—that is, proving that space beyond Mars and the asteroid belt were hardly impenetrable and no special obstacle to the universe beyond—was the first barrier. And it was an important one, because if Pioneer had shown the belt to be dangerous, the implications for future exploration would have been profound, necessitating either the development of immensely powerful rocket boosters that could lift a spacecraft around the belt or the invention of some means of protecting a craft from the onslaught of asteroid particles.

Even without the asteroid belt, there would be no hope of exploring the outer solar system without the technological wherewithal to build and operate a spacecraft able to function for the years necessary to make the journey. Pioneer amply demonstrated that it was possible not only to do so but to do so cheaply and simply, even if future missions chose to spend more money. Electronic circuits and scientific instruments could be designed and built that could withstand the cold and radiation of space for not just a couple of years but indefinitely. Radioisotope thermoelectric generators weren't a gimmick or an empty promise; they proved to be an efficient and reliable means of powering a probe far from the Sun. Trajectories could be devised and computed that could take a spacecraft hundreds of millions of miles to a rendezvous timed to within seconds of the actual encounter. And even across such immensities of empty space, it was possible to stay in touch and receive priceless data.

Perhaps most importantly, *Pioneers 10* and *11* transformed the outer solar system from a distant realm ruled largely by speculation and uncertainty to a tangible destination, to be visited, directly examined, and ultimately understood by humanity's robotic representatives. For the first time, Jupiter and Saturn had become real places that had been measured, studied, and seen close up. Theory had been supplanted by fact and old questions had been answered, if only to be replaced by new questions that no one had ever thought of before.

Unable to directly touch the objects they study, astronomers plumb the secrets of the stars and the universe through the light and other forms of electromagnetic radiation that heavenly bodies emit. Expanding the frontiers of knowledge means improving their instruments by building larger telescopes and other, more sensitive tools. But even the largest telescope in the world is immersed within Earth's obscuring atmosphere, and there are physical and practical limits to the size and sensitivity of any scientific instrument. It's only been within the past 50 years that another option has become available: taking instruments outside of Earth's environment into space. Scientists such as James Van Allen and John Simpson led the way by placing their experiments on high-altitude balloons and rockets, then on orbiting satellites. Finally, instead of trying to bring the universe closer to Earth, they could move their instruments a little closer to the universe.

Pioneer and the other deep-space probes were the culmination of that process. If astronomers could never attain the ideal vantage point for their studies—going there in person—they could send mechanical representatives in their place. Ultimately it was the best solution because spacecraft can go where humans would never survive, as Pioneer did when plunging into the radiation belts of Jupiter.

For thousands of years, science could only be done by direct human observation. Our eyes had to see for themselves, our hands reach out and touch for themselves. With spacecraft such as Pioneer, the extension of our human senses that began with the telescope became a means to extend not just our senses but also our minds and presence to places unreachable by our physical selves. Even if no human being has yet literally visited Jupiter or Saturn, it is not inaccurate to claim that humans have already been there through the spacecraft that represented all of humanity.

10

Planet of Clouds

For a planet that had been visited so often by American and Russian probes in the 1960s and 1970s, Venus still held many unanswered questions enshrouded within its opaque atmosphere. In many ways Venus seemed to be an Earth gone terribly wrong. A runaway greenhouse effect, the phenomenon in which gases such as carbon dioxide and water vapor trap heat within a planet's atmosphere, had apparently driven the Venusian surface temperature to 900 degrees Fahrenheit, hot enough to melt lead. There was no water, no oxygen, certainly no life. How had such extreme conditions arisen on a world so similar to Earth in size and mass that it was considered our sister planet? What were the implications of the Venus environment for the fate of Earth?

Venus had been a target for spacecraft from the earliest days of spaceflight. As the planet closest to Earth, it was a natural destination for scientists and engineers taking the first toddling steps toward deep-space missions. Humankind's first successful interplanetary mission, *Mariner 2*, flew past Venus in 1962. The Russians practically bombarded Venus with spacecraft in the 1960s, including some probes that impacted the surface and managed to return some data before succumbing to the high temperature and pressure. Other American missions provided only hints at what might lie beneath the Venusian cloud layers. Even by the beginning of the 1970s, more was known about Mars and Jupiter than about Earth's much closer sister planet.

A detailed examination of Venus, one that would do more than give the planet a brief passing glance, was definitely overdue. An orbiting spacecraft could make long-term observations of cloud patterns and other phenomena and, using radar, penetrate the cloud layers to map the surface. Atmospheric probes, more durable than those built by the Soviets, could analyze the structure and composition of the Venusian atmosphere. Such ideas had been part of mission proposals and studies for years and by 1970 had been crystallized into a Space Science Board report entitled *Venus—Strategy for Exploration*, referred to as the Purple Book because of the color of its cover. The Purple Book had grown out of an earlier Goddard study on a Venus mission, spearheaded by Richard Goody, a Harvard atmospheric physicist, and Donald Hunten from the University of Arizona.

When Russia's *Venera 7* became the first spacecraft to send data back from the surface of another planet in December 1970, the momentum for another American Venus mission picked up. Much to the disappointment of the Goddard team that had originally planned the mission, the deputy director of Goddard and others at NASA Headquarters decided that Ames Research Center was better suited to handle the job, not only because of its record with Pioneer but also thanks to its research on planetary atmospheres and atmospheric probes. A 1971 Ames experiment called PAET, the Planetary Atmosphere Experiments Test, sent a prototype probe into space on a rocket and brought it hurtling back into Earth's atmosphere as onboard instruments took measurements and samples. PAET was a spectacular success that laid the groundwork for all the atmospheric planetary probes that would follow—not to mention proving that Ames knew how to send spacecraft into the Venusian atmosphere.

A little campaigning also helped. "John Foster and Skip Nunamaker pushed for Ames to do the job," Charlie Hall recalled. "Those guys were on that damn airplane from Ames to headquarters and back two or three times a week. I didn't see how they could take it." Foster and Nunamaker had worked with Hall for years, and not only were they well aware of what the Pioneer team could accomplish, they were most persuasive in convincing any doubters. In January 1972, NASA gave the Venus orbiter/probe mission to Ames, where it became *Pioneer Venus*, the final mission to be developed and managed by the Pioneer Project Office.

BACK TO THE NEIGHBORHOOD

Although it wouldn't be making the epic journeys of the previous *Pioneer 10* and *11* craft, with regard to either distance or time, *Pioneer Venus* was hardly going to be a walk in the park. In fact, in many important ways it was actually far more complex than the Jupiter missions. Instead of one spacecraft, *Pioneer Venus* would consist of two main craft: the orbiter and a multiprobe bus that would carry probes to be launched into the Venusian atmosphere.

The engineering challenges might be vastly different than those of a deep-space mission to the outer solar system, but they were no less daunting. True, power wouldn't be a problem, as solar cell arrays and batteries would work perfectly well this close to the Sun. And there would be no worries about the asteroid belt or intense radiation at Venus. But how to build probes that could survive long enough to send back their data before being destroyed by the furnace-like heat, enormous pressures, and sulfuric acid-laden clouds of Venus?

Coordinating and communicating with multiple spacecraft simultaneously would also be an entirely new challenge for the Deep Space Network. And with two large spacecraft and several probes, there would be many more scientific instruments and experiments aboard Pioneer Venus than the previous missions, and all had to not only physically fit into the spacecraft but work together as part of an integrated system. Most challenging of all was the requirement that this be a low-cost mission.

Two main bidders stepped up to take on the job: long-time Pioneer contractor TRW and Hughes Aircraft. NASA gave study contracts to both manufacturers to develop detailed proposals and then would select the winner. "The technical evaluation of the Hughes and TRW proposals was pretty close," Charlie Hall said. The proposals were also extremely detailed, something that caused a slight problem when Hall and John Foster went to NASA headquarters to present them for the final selection. "We had enough documents that we took a 3 × 4 wagon stacked about 4 feet high," Hall recalled. "We wheeled the wagon from our car in the garage into the elevator on the ground floor. We found that the thing was so big we couldn't get into the elevator. After a few seconds the elevator doors close. Here we are with all this confidential information in the elevator, and we don't know where the elevator's going. So poor John Foster starts running up the stairs, seeing if he can get to each floor and stop the elevator. I stayed on the ground floor and kept pressing the down button. Finally I could see it had reached the top and started down. The doors opened and everything was still there. I was sure that when those doors opened, the thing would be empty. I don't know what we would have done!"

After rolling the wagon out of the elevator so that one man could get in first, Hall and Foster managed to get upstairs and get to work. "John Foster made the presentation that day to George Low," said Hall. It was a close competition. "Several of the components that TRW proposed for *Pioneer Venus* were spares from *Pioneer 10* and *11*. Which was okay, but then somehow TRW and Martin Marietta were teaming up on the probes, with Martin building the probes. This is something like the tail wagging the dog, because Martin would have a bigger share of the total project, but TRW would be the managers."

Such an arrangement would have complicated the project's management and possibly contributed to budget overruns.

In February 1974, NASA chose Hughes because of its simpler spacecraft design and less complicated management proposal. Robert Kraemer of NASA headquarters noted that Hughes was also attractive because they didn't see the *Pioneer Venus* contract as a profit-making venture—"Hughes would be content to break even." Instead, the project would bring other benefits to the company. Bud Wheelon, head of the Hughes space division, told Kraemer that although they already made more than enough profit from their satellite business, "his engineers were getting bored with designing variations of the same old communications satellites. He was beginning to lose key people and he wanted the *Pioneer Venus* work because it was technically challenging and exciting and would rejuvenate his work force."

Amid some budget problems that pushed the launch of the mission to 1978, *Pioneer Venus* began to take shape. The two main craft, the orbiter and the multiprobe bus, would use the same basic configuration. The spacecraft would be launched separately, first the orbiter and then the multiprobe, arriving at Venus a few days after the orbiter. The bus would carry four probes, one large and three small, each packed with instruments. As the multiprobe approached Venus, it would release the four probes to penetrate the planet's atmosphere at widely separated points. The three small probes would be protected by heat shields so that they could take measurements until impacting the Venus surface; the large probe's descent would be slowed by parachute. The bus itself, also carrying some instruments, would plunge into the atmosphere and make some observations before burning up. Meanwhile the orbiter would observe Venus from above, mapping the surface by radar and taking other measurements.

NASA had never attempted to build anything quite like the probes before. They would be four tiny self-contained spacecraft that could handle temperatures up to 900 degrees Fahrenheit, atmospheric pressures 100 times greater than Earth sea level, and immersion in

Venus and its visitors: the *Pioneer Venus* multiprobes and bus (left) and the orbiter.

sulfuric acid rain, all while hurtling toward a planet's surface at over 26,000 miles per hour. Each probe was a titanium sphere encased in a blunt forward heat shield and an aft shield; inside the sphere the science instruments were tightly packed. For some of the instruments to take their measurements, however, they couldn't be completely sealed inside a metal sphere. Some sort of window or aperture to the outside environment had to be provided, but how to do so without exposing the rest of the craft's interior to the hostile conditions outside?

The question was complicated by the fact that no single window material would work for all the instruments. For the experiments that observed in visible light and ultraviolet, industrial-grade sapphire would do the job. But not for the infrared radiometer aboard the large probe, because sapphire would be opaque to infrared. "The only suitable material for that window was diamond, and it had to be a huge three-quarters of an inch in diameter," recalled Robert Kraemer.

"Oh boy, I thought, wait until the news media picks this up. We will be a sure candidate for Senator Proxmire's infamous Golden Fleece Awards." But thanks to a bit of Pioneer luck, a lower-cost type of slightly impure industrial-grade diamond was found to be suitable, and the project commissioned a dealer to locate two specimens of suitable size in South Africa. "In fact, we were able to avoid $12,000 in U.S. import taxes by arguing that the diamonds were only 'in transit' from the United States to Venus," Kraemer related.

Not all the hurdles were overcome quite so easily. "It went on and on, all these new things," Hall said. Perhaps the most intransigent and exasperating problem was the development of the parachute for the large probe. "We knew pretty much that the atmosphere of Venus is highly acidic. So we would test parachute material in an acid bath like we thought would be encountered on Venus. The damn stuff would end up in shreds." Obviously, designing a parachute that could survive sulfuric acid rain and the pressures and wind speeds it would encounter as it deployed in the Venusian atmosphere wasn't going to be easy. Engineers used one of the largest enclosed structures in the world, the Vehicle Assembly Building at Kennedy Space Center, to conduct drop tests. They tested parachute deployment in the Ames 40 × 80 foot wind tunnel—the chute wouldn't open. "We'd put a parachute in a canister on the bottom of an F-104," Hall said, describing still another test. "The F-104 would get up there and dive down to get a high speed and then eject the parachute from the canister, and we'd see what happened. We had motion pictures, 100 frames a second. This damn parachute came out of the canister, and in one frame everything was okay. The next frame, 100th of a second later, the thing was in shreds."

More modifications and tests followed, with anxiety increasing as the mission launch window rapidly approached. With time growing short, some tests had to be combined in a balloon launch from White Sands, New Mexico. A model probe was lifted to 19 miles and dropped, testing not only the parachute but the release of the probe heat shield and the separation of the spherical pressure vessel. But the test was a miserable failure, with the parachute breaking and the probe model hurtling to the ground. It was beginning to look as though *Pioneer Venus* was not going to get its workable parachute. Fortunately for the project, though, further analysis of the test proved that the culprit hadn't been the parachute but rather a mechanical mishap that caused the probe to tumble erratically so that the chute released at the wrong time and the wrong angle. The problem was corrected and the balloon drop test repeated, this time with complete success. "Fortunately, Charlie Hall's hair was already white, so we saw no permanent impact on him from this adventure," Robert Kraemer later wrote.

If *Pioneer Venus* was Charlie Hall's management swan song, it also proved to be his most difficult challenge, the culmination of all he had done and learned on the previous missions. Aside from the herculean task of building and managing two separate spacecraft and their missions at once (actually, six separate craft, counting the probes), he had to coordinate, manage, negotiate with, and placate a team of principal investigators over twice the size of the *Pioneer 10* and *11* team. With 30 separate instruments and experiments on *Pioneer Venus*, reaching the often necessary compromises and consensus took consummate political skill. And unlike the previous missions that for the most part enjoyed collegial and friendly relationships among the PIs, *Pioneer Venus* saw some not-so-friendly rivalries and conflicts. "There were some scientists [who] would criticize other scientists—they don't know what they're doing, that's not important what they're doing—that sort of thing," Richard Fimmel diplomatically explained. "There's always someone who thinks he knows more than the other guy."

There were also the occasional clashes between project personnel and the contractor. "We had our differences," Hall recalled with a laugh. "I don't want you to think that this was a Sunday afternoon picnic every time we got together. We'd yell and scream and everything else at some of these meetings." Sometimes a different sort of tactic was called for. "I had the philosophy that I never wanted to call my boss to help me with a problem. You solve it first. Don't drag your boss into it." When Hughes fell far behind schedule in software development for *Pioneer Venus*, Hall realized he needed to break that rule for once. "We called a meeting and I thought, this is the one time I'm gonna need all the push I can get, because this is key—if we fail here, the whole mission fails. So I asked Hans [Hans Mark, director of Ames] if he wanted to go with me. And we met with their equals on the Hughes team. But we won. It so happened that Bud Wheelon and Hans Mark were classmates at MIT." Not that such a personal connection made things too easy. "That was a knockdown meeting, no punches held, but at the end of the meeting, we all shook hands and knew we were on the right track and kept going. Then you say, 'Let's go out and have a drink,' and we could revert to being humans again."

With two separate launches, another issue Hall had to settle was the choice of boosters. Some within NASA pushed for the use of two different rockets, but Hall fought the proposal. "I said it was an awful idea. They're proposing to save about 3 million bucks by going with two different launch vehicles. I could spend 3 million bucks just on managing those two things. With a contract where we have a single launch vehicle, everyone gets to know each other and we manage just one deal. If I have to manage two, I'll need 10 more people." Hall was right. going with the more expensive Atlas Centaur for both launches ultimately saved money by keeping total costs down.

By early 1978 both spacecraft were finished and shipped to the Cape for final preparations. First the orbiter would leave Earth, becoming *Pioneer 12*; then the multiprobe would become the last of the Pioneer line, *Pioneer 13*. The Orbiter would take a longer, slower trajectory, so that it would arrive at Venus at a lower velocity and thus require less braking thrust—and consequently, needed to carry less propellant onboard—to slow itself to enter into orbit. The Multiprobe would take a faster, shorter course, releasing its four probes 20 days before it reached Venus itself. Both trajectories would be planned and timed so that the atmospheric probes would arrive at Venus only a few days after the Orbiter. Like all the other Pioneer craft, all of the *Pioneer Venus* spacecraft were spin stabilized.

It was a complex proposition, to be sure. Timing and coordination would be paramount. It was true that with Venus much closer to Earth than Jupiter or Saturn, the light-speed communications delay was much less, so there would be no long, agonizing radio blackouts or waits to find out whether a maneuver had happened successfully. But that also meant that when everything happened, it was going to do so all at once, with little time to fix glitches or work through problems. If disaster occurred, controllers would know within minutes, not hours. While all the Pioneer missions were real-time because they were flown from the ground, Pioneer Venus would be the most "real-time" of them all.

A LONG SHORT TRIP

The *Pioneer Venus* orbiter began its journey on the morning of May 20, 1978. The launch was right on schedule, without any of the all too typical last-minute delays that usually plagued a rocket launch, and the craft became *Pioneer 12* upon its departure from Earth. The multiprobe took a little longer to become *Pioneer 13*, its scheduled launch of August 6 delayed due to some ground equipment problems until early in the morning of August 8. But the *Pioneer Venus* mission was now fully under way. In less than 6 months the main phase of the mission would be over.

The passages of both craft to Venus were quiet and, aside from several minor and correctable glitches, without any alarms or emergencies. Along the way the orbiter picked up several extremely strong gamma ray bursts, an unusual phenomenon that had only been first observed by astronomers in 1973. The orbiter's data were correlated with simultaneous observations of the bursts from several other spacecraft in Earth and solar orbit.

Meanwhile the Pioneer controllers were preparing the multiprobe for its big moment: release of its four probes. One spacecraft was going to become five separate and independent craft, each with a different destination. The preparations had to be completed perfectly because once the probes were launched from the multiprobe bus, they were on their own—they couldn't be controlled from Earth. Each battery-powered probe would remain inactive until it entered the Venusian atmosphere, and each was equipped with a timer that would activate its systems at the proper moment. If a probe were turned on too early, it might run out of battery power before reaching Venus, becoming a useless projectile; if turned on too late, it would miss crucial data during its initial entry into the atmosphere. The timers had to be programmed before the probes were released, which demanded precise calculations of the transit time of each probe from release time to Venus arrival.

The large probe, centrally mounted on the multiprobe, was the first to go, on November 16. Four days later Pioneer controllers oriented the spinning multiprobe to the proper position and released the three small probes, the timing of each release calculated to send each individual probe on its own path to Venus. The timers started clicking away the seconds and all four probes were alone in space, silent and out of touch with Earth as they hurtled toward their destinations. The course of the multiprobe bus, now without its four traveling companions, was adjusted so that the craft would follow the probes into the Venusian clouds.

Once again, December seemed to be Pioneer's month. On December 4, 1978, 5 years and 1 day after *Pioneer 10* made its historic Jupiter flyby, the *Pioneer Venus* orbiter prepared to enter orbit around the second planet. After the excitement of the probe release, the orbiter's arrival provided a few more white-knuckle moments. The braking engine was a solid-fuel rocket simple, lightweight, inexpensive. It would be used only once, so there was no need for a more complex liquid-fuel engine. But no solid-fuel engine had ever before been fired after spending so long in the cold of outer space. Would it ignite when the time came? If not, the orbiter would sail right past Venus into solar orbit. And since solid-fuel rockets aren't control-

lable—once ignited, they simply burn until their fuel is exhausted—
the orbital insertion burn had to be timed just right.

To make matters worse, the engine burn would take place while
the orbiter was behind Venus and out of communication with Earth.
Pioneer controllers had to program the firing sequence into the
orbiter's command memory before the craft disappeared behind the
planet, but they wouldn't know if it had worked until the orbiter
emerged. To hedge their bets, they decided to transmit a second firing
command when the orbiter came back into view, just in case the engine hadn't fired behind Venus. The orbit resulting from such a late
engine burn wouldn't be quite what the controllers and PIs wanted,
but at least it would prevent the orbiter from missing Venus entirely.

After slightly less than half an hour of radio silence, the orbiter
emerged from behind Venus at 8:14 Ames time on the morning of
December 4. The Pioneer Mission Operations Center quickly ana-
lyzed the signals from the craft. The engine burn had been successful.
The orbiter had earned its name and had become a satellite of Venus.
Instrument checkouts, calibrations, and tweaking of the orbit ensued.

Five days later the most dramatic part of the mission began as the
probes approached entry. The timers aboard each small craft turned
each on and they reestablished contact with Earth. In about 20 minutes, all four probes would plunge into the atmosphere and the DSN
would take center stage. "The big challenge of the mission was to pick
up the radio signals from the probes after the entry," Charlie Hall
recalled. "When you first hit the atmosphere, it ionizes and cuts off all
radio communication. So the problem was to know precisely where
you're looking and see if you could pick up these small probes going
into the atmosphere."

Detecting the probes after the ionization blackout turned into a
friendly game of one-upmanship between the DSN station in
Canberra, Australia, and the one in Goldstone, California, the two
stations in view of Venus at probe entry time. "They're competitive as
hell," said Hall. "Who's gonna pick up those signals first? We'd gone
through the ionization and all of a sudden from Australia: 'We got

An image from the *Pioneer Venus* orbiter.

'cm!' It was less than 3 seconds later that the guy from Goldstone says, 'We got 'em too!' Each side was really hoping to be the first one. That was a fantastic communications feat because the signals were so weak, and to pick these up, all four of them in such a short time, and lock on to the signal and get data back, it was a challenge that those JPL guys met."

But how long would each probe survive in the hellish environment of Venus? No one was expecting much, of course. The probes had only been designed to withstand the dive to the surface and transmit data during that brief period, not beyond—and yet.

"As these three probes descended, we had predicted precisely how long before that spacecraft would hit the ground and quit transmit-

One of the multiprobes on the scorching surface of Venus.

ting," Fred Wirth said. "I remember sitting there watching the data come in on the three probes and the fourth probe with a parachute. And one by one, okay, probe one crapped out, okay, probe number two crapped out, that's it. And people started leaving. Finally the last one quit and everybody started walking out. Okay, it's the end of the multiprobes."

It wasn't a surprise or disappointment, just the expected end of the probe mission—almost. "And all of a sudden I saw on my monitor there was data coming in," said Wirth. "I thought, where the hell is this data coming from? It was probe data. Initially I thought JPL was playing back a tape or something, but I found out it was in real time. I jumped up and ran down the hall, 'Get back in here! Everybody get back in here! One of the probes is still working.' And I'll be damned, it lasted for another hour and 5 minutes, I believe, more than an hour after it had impacted Venus. One scientist says, 'The only thing I can think of is it must have landed in a sand trap with the antenna pointing straight up.' So he and I were sitting there watching the data displaying the internal temperature of his instruments, and we thought, 'My God, how hot is it gonna get?' The temperature started climbing and climbing. We thought, 'How long it is gonna last before the spacecraft craps out?' We started taking bets. Sure enough, it was more than an hour before it conked out. But it was one of those things where everybody said, 'Oh yeah, this is all over with.'" One of the small probes survived and continued to send back data for 67 minutes after impacting on the day side of Venus in the southern hemisphere.

About an hour and a half after the last probe entered the atmosphere, the empty multiprobe bus followed and burned up, its two onboard instruments sending back data until the end. All four probes had performed spectacularly, opening new windows on the Venusian atmosphere despite their brief lives. John Noble Wilford of the *New York Times* summed it up elegantly: "A rain of science fell on Venus today," he wrote. Now the *Pioneer Venus* orbiter would settle into a long period of observation and discovery. And once more, a Pioneer spacecraft would exceed all expectations.

THE NEW VENUS

Perhaps more than any of the previous Pioneer missions, *Pioneer Venus* was the least glamorous as far as the public was concerned. The mission didn't have the upfront and easily understandable appeal of *Pioneer 10*, going someplace for the first time, seeing things close up that humans had never seen before, facing totally unknown dangers. It didn't promise gorgeous pictures of a distant world: Venus might be pretty in the morning or evening sky, but it was a boring, featureless sight through a telescope. All it offered were colorless clouds. There weren't even any interesting moons to explore and no chance of finding any, since they would have already been seen with Venus so

close to Earth. Spacecraft had already been there anyway, so what was
the big deal? If Venus had been the lush, tropical jungle world of previous fanciful speculations, another close-up look would be interesting. Instead, it was just a planetary blast furnace, completely lifeless.

But although Ames may not have been quite as inundated by the
press for *Pioneer Venus* as it was for *Pioneers 10* and *11*, and although
television stations didn't interrupt their programs to broadcast pictures from the *Pioneer Venus* orbiter as they had for pictures of Jupiter, the mission was of supreme interest to the scientific community.
Far from being just another planetary mission, *Pioneer Venus* was
nothing less than the first full-scale survey of another planet. Aside
from the volumes of new information it provided about Venus, it
served as a template for later similarly ambitious missions to other
planets.

The probes provided the first surprises with their data on the
atmospheric composition of Venus. They found that the chief component of the Venusian atmosphere is carbon dioxide, as expected,
with clouds formed mainly of sulfuric acid droplets. Unexpectedly,
the probes also detected large amounts of argon and neon, hundreds
of times more than in Earth's atmosphere. The argon and neon were
primordial in form, meaning that they had been present since the
formation of the planet instead of arising later from the radioactive
decay of other elements. Scientists had assumed that Venus and Earth
had probably been formed from the same cloud of primal gas at the
origin of the solar system. But if the amount of primordial elements
on each world differed so radically, those assumptions were wrong
somehow. The discovery led to some radical rethinking of the origins
of the solar system, a revised theory that could account for the differ-
ing levels of primordial elements among the inner planets. If this
wasn't enough, the data gathered by the four separate probes, along
with the multiprobe bus, made it possible to assemble a planet-wide
picture of the temperature, density, composition, and structure of
the Venusian atmosphere.

Even after the drama of the probes' entry, the *Pioneer Venus* mission was far from over. The orbiter began a detailed reconnaissance of Venus, using its radar imager to provide the first real look at the surface and topography of the planet. Venus turned out to be fairly uniform in geography, largely without the high mountainous regions or deep canyons of Earth and Mars. The highest and lowest points on Venus describe a much narrower range, so that the planet is relatively smoother than the other rocky worlds. Also unlike the other terrestrial planets, Venus is nearly a perfect sphere; it lacks the polar flattening and equatorial bulges those worlds exhibit. These findings seemed to indicate that Venus might lack the plate tectonic activity so vital to Earth's own geography. Not that Venus is completely inert geologically; Pioneer also found indications of possible active volcanoes and areas of surface uplifting and downwelling. But the geological contrasts between Venus and Earth provide a major clue as to why each world has found its particular fate: Earth warm, wet, and life giving, and Venus hot, dry, and sterile.

In keeping with the mythological origins of the name of Venus, scientists named the planet's surface features exclusively after women, both mythological (Aphrodite, Ishtar, Freya, Hesperos) and real (the Egyptian queen Cleopatra, the writer Colette, the physicist Lise Meitner). The International Astronomical Union, the body charged with naming all newly discovered astronomical bodies and features, made it all official. Computer-generated three-dimensional maps from Pioneer data provided the IAU naming committees with even more opportunities to exercise their mythological and historical knowledge. As the geography of Venus was mapped in still more detail by later probes such as *Magellan*, the nomenclature followed the same scheme.

The orbiter also spent its years of work observing the Venusian cloud systems in ultraviolet light, tracking cloud movement and speed, variations in density, and other features such as the effect of the Sun on the atmospheric patterns. Pioneer verified the greenhouse effect at work on Venus, its clouds trapping heat from the Sun and distributing it all across the surface. And although not seeing it di-

rectly, the orbiter detected evidence of lightning occurring in the atmosphere. Without any magnetic field, Venus is affected by the solar
wind much differently than planets with a strong magnetosphere such
as Earth or Jupiter, and the orbiter also studied these interactions.

The *Pioneer Venus* orbiter also found time to turn its attentions
away from Venus on occasion. Most importantly, it made up for the
lack of any American mission to Comet Halley in 1986 (thanks to the
usual NASA budget woes) by turning its instruments on the comet
during its swing around the Sun. Realizing the importance of such
observations and the opportunity that *Pioneer Venus* provided, the
PIs agreed to temporarily suspend the orbiter's primary mission of
studying Venus for several months as Comet Halley passed through
the inner solar system. A bit of fancy maneuvering was necessary to
direct the orbiter's view away from Venus and keep its instruments
focused on the comet, but the results were worth it. Unlike the other
international missions that could only study the comet on its way to
or from the Sun, the orbiter's unique vantage point at Venus allowed
it to make observations close to perihelion, the comet's closest passage to the Sun. It watched as the comet's icy core began to melt close
to the Sun and then refroze as it headed back to the outer solar
system. The orbiter also managed to briefly investigate several
other comets during its mission, supplementing the data from other
spacecraft.

In the grand Pioneer tradition, the orbiter continued operating
long past its predicted lifetime of 243 days, and delighted scientists
made the most of it. After 10 years in orbit, the spacecraft was showing a few signs of age, including some degradation of its solar cells
resulting in a loss of power and the malfunction or failure of some
instruments and spacecraft systems. But nothing was critical, and the
Pioneer controllers and engineers managed to work around most of
the glitches that arose. The spacecraft's orbit was adjusted at various
times to optimize different observations, using its maneuvering
thrusters.

Finally, by October 2, 1992, the maneuvering fuel was exhausted, and the *Pioneer Venus* orbiter began to descend into the atmosphere as its orbit decayed. But even as the orbiter followed its sister craft to the surface at last, it continued to transmit data until it disappeared behind Venus. Controllers hoped for one final signal, hoping that *Pioneer Venus* might yet emerge alive for just a little longer, but it was not to be. The orbiter burned up and was never heard from again, after spending 14 years making 5,055 orbits of Venus.

In August 1990 the *Magellan* spacecraft followed Pioneer Venus into orbit. As its predecessor died, it picked up where Pioneer Venus left off, going on to map Venus in even greater detail and resolution. Once again, Pioneer had blazed the trail for others to follow.

THE LAST OF THE LINE

Not only was *Pioneer Venus* the last of the Pioneer series, a noble and honorable distinction, but for 11 years it also represented the last planetary space mission launched by the United States. After leading the exploration of the solar system for most of the space age, showing humanity sights it had never seen before, America shrunk back, stopped looking outward, and turned its eyes inward.

It wasn't just in planetary exploration but in the idea of exploration in general that America's ambition waned. The focus turned from exploration to exploitation, from finding out new things to making space pay. The object of all these hopes was the Space Shuttle, to be joined shortly by an orbital space station. The reality fell far short of the hopes, however. The Space Shuttle turned out to be neither cheap nor particularly safe, nor was it necessarily well suited to every conceivable space mission, despite efforts to make it the be-all and end-all of space transportation. Plans for a space station became an administrative, financial, and engineering nightmare. Yet NASA had committed itself so heavily to and invested so much of its future in these two projects that there was little money or other resources left over for anything else, particularly pure science missions.

When the agency decided to dispense with expendable booster

rockets entirely and use the Space Shuttle as its sole launch vehicle, some projects already in the works, such as *Galileo*, had to go back to the drawing board and revamp themselves to conform to entirely new requirements, driving up their expenses even further and unwillingly turning themselves into irresistible targets for congressional cancellations. After the loss of *Challenger* in 1986 and the suspension of Space Shuttle flights that followed, NASA was left with no means of putting up even a small Earth satellite, much less a mission to another planet.

Slowly, with the launch of *Magellan* and *Galileo* in 1989 (both via the Shuttle), NASA returned to the solar system. NASA administrator Daniel Goldin, who took office in 1992, espoused a "faster, better, cheaper" philosophy that owed a lot to Pioneer's example—as a former TRW executive, Goldin was well aware of the project's success. Unfortunately, the "faster, better, cheaper" notion often tended to follow the spirit but not the techniques of Pioneer and led to some embarrassing and expensive failed missions in the 1990s.

Pioneer Venus was the last of a special breed. When the orbiter burned up in 1992, it marked more than just the end of a single mission or series of spacecraft. It was the coda for humanity's first, most exciting, and most adventurous era of planetary exploration. More missions would follow, and many would be much more ambitious and sophisticated. But none would ever again have that feeling of being *first*, of facing complete unknowns, of the surprise and discovery that come from knowing that anything is possible.

None again would be pioneers.

11

Whispers Across the Abyss

t was all over. Or at least so it seemed. As 1980 began, Jupiter and Saturn had been left far behind, and although both *Pioneers 10* and *11* were still cruising along, their primary missions were completed. The *Pioneer Venus* probes had done their jobs and were now so much detritus on the furnace-hot surface of Venus. The *Pioneer Venus* orbiter continued to circle the planet, mapping and observing, but aside from the routine housekeeping functions of monitoring its systems and receiving its data, it didn't require any special attention. Nor did the remaining solar probes, *Pioneers 6, 7,* and *8. Pioneer 9* had finally given up the ghost in 1983, but the others continued in their stately orbits around the Sun, still serving as solar weather sentinels and observation outposts whenever Pioneer controllers had the chance to contact them.

But there would be no more Pioneer missions. *Pioneer Venus*, also known as *Pioneers 12* and *13*, would be the last. New missions would be planned to return to Venus, Jupiter, and Saturn, and Ames Research Center would even participate in some of these projects, some of which would be direct descendants of Pioneer concepts. But no more spacecraft would bear the Pioneer name.

After 38 years working for NASA—18 years of those, nearly half his career, heading the Pioneer project—Charles F. Hall decided to retire in March, 1980. He received NASA's Distinguished Service Medal, the agency's highest honor, along with a wing-ding of a retirement party thrown by his Pioneer and Ames colleagues. He would go on to settle into a quiet retirement not far from Ames in Los Altos with his wife Constance.

The Pioneer project office at Ames was also officially closed at around the same time, its work finished, with no new Pioneer missions to develop and organize. The operation and management of the remaining Pioneer spacecraft were transferred to the space missions branch under the direction of Richard Fimmel, who had already served Charlie Hall and Pioneer well for many years, most recently as project science chief.

It may all have seemed rather elegiac and somber to those on the outside. It was anything but. Although NASA might have relegated Pioneer to the historical and administrative dustpile, the spacecraft themselves appeared to have other ideas, stubbornly refusing to accept official proclamations of their demise even as the humans who built and launched them moved on.

EXTENSIONS

For some of the Pioneer principal investigators, the missions were essentially over after the planetary encounters. Tom Gehrels's imaging photo polarimeter, for example, didn't have much to look at after *Pioneer 10* left Jupiter and *Pioneer 11* left Saturn. With the spacecraft's Sun sensor losing effectiveness as the distance to the Sun increased, the IPP took over its role to gauge the rotation and orientation of the Pioneer craft by sighting on bright stars and also helping to calibrate the particles and fields instruments by providing what became known as the "Doose correction," after Lyn Doose, the IPP team member who devised the technique. Other instruments were inoperative or turned off to save power.

For the particles and fields investigators, though, the mission continued. Van Allen's Geiger tube telescope, Simpson's charged particle experiment, and Frank McDonald's cosmic ray telescope were free of the intense and overpowering radiation flux of the planets, but they still had a lot to do. Now well past the interference of the planetary magnetospheres, they were searching for the edge of the solar system itself, the physical boundary that separates the Sun's realm of influence from the rest of the galaxy.

The casual observer might consider the obvious outer marker of the solar system to be the orbit of Pluto, the farthest planet from the Sun. But that's only one possible definition and one that discounts the most important member of the solar system, the Sun itself. For astronomers, defining the true edge of the solar system means finding the region where the sun no longer holds sway, where the solar wind dwindles and vanishes, superseded by the interstellar medium. This is called the heliopause, and with the conclusion of the planetary encounters of *Pioneers 10* and *11*, the heliopause became the new, if rather more ephemeral, destination of the two spacecraft.

Ever since the discovery of the solar wind, the flow of charged particles from the Sun, debate had raged among scientists about just how far out it extended from the Sun. Obviously it could be felt at least as far as Earth's orbit, since the influence of the Sun on Earth was well documented. But how much farther did it go?

"In the early part of the mission the solar wind limit was thought to be just beyond the orbit of Jupiter," Van Allen said. *Pioneer 10* quickly disproved that notion. The extent and intensity of the solar wind could be gauged by the way in which it affected, or modulated, the flux of cosmic rays coming from outside the solar system. Basically, the more galactic cosmic rays that were detected, the weaker the solar wind, and vice versa. The ratio between the two factors is called the galactic cosmic ray gradient, and it was expected to increase sharply after *Pioneer 10* left Jupiter. It didn't, much to the surprise of Van Allen and the other particles and fields PIs. The influence of the Sun was felt much farther than previously expected.

As *Pioneer 11* followed *Pioneer 10* toward interstellar space, it confirmed the findings. "The farther out Pioneer's gone, we had to keep
going back to the drawing board and keep revising the estimates,"
Van Allen explained. "It's kept ahead of us all the way like a rabbit on
a track." The extent of the heliosphere appeared to be tied to the 11-
year solar cycle, which meant that it would expand and contract as
the Sun's activity increased and decreased. If so, one or both of the
Pioneer spacecraft should eventually hit the heliopause, either by
catching up to it or as the heliopause itself passed the spacecraft as it
contracted toward the Sun.

The odds that one or both probes would find the heliopause were
increased by the fact that each was moving outward in a different
direction. Just as the Earth orbits around the Sun, the Sun orbits
around the center of the Milky Way galaxy, and the Sun's magnetosphere is believed to be blunted in the direction in which the Sun is
moving, just as those of Earth and other magnetospheric planets inside the solar system. *Pioneer 10* was moving out into space in the
direction opposite the Sun's orbital motion, presumably through a
tail of solar wind that might extend billions of miles. *Pioneer 11* was
heading into the galaxy in generally the same direction as the Sun,
and would thus pass through the Sun's bow shock from the inside
and out into the interstellar medium more quickly than *Pioneer 10*.
Scientists were pretty certain of their general heliospheric model, but
which spacecraft would win the race to true interstellar space was an
open question.

Unfortunately, it was a question less dependent on timing than
on practical considerations. Even if the *Pioneer 10* and *11* missions
were officially still going concerns according to NASA, their "extended
mission" status put them at relatively low priority as far as the precious allocation of Deep Space Network tracking time was concerned.
New missions almost always came first, not older missions that had
already fulfilled their primary objectives. For the Pioneer PIs searching for the heliopause, this simple, hard truth became evident even as
early as 1976. The Viking missions to Mars were demanding vast

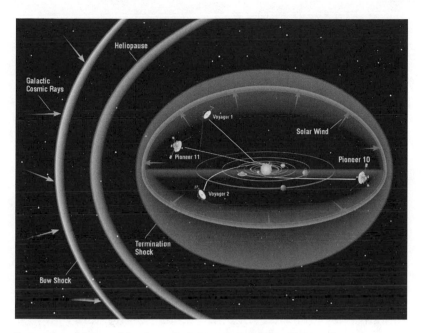

Four spacecraft bound for the end of the solar system and interstellar space.

amounts of DSN resources, and Pioneer tracking allocations were consequently cut. In a joint letter to Ichtiaque Rasool, NASA deputy administrator for space science, Drs. Van Allen, Simpson, and McDonald made their case: "*Pioneer 10* has moved past the orbit of Saturn and the experiments onboard are beginning the first in situ exploration of the outer heliosphere. . . . The *Pioneer 10* mission offers the only prospective opportunity to explore it during a period of near solar minimum activity. . . . It is our strong belief that increased coverage will be very rewarding in new scientific findings. Indeed, failure to provide adequate coverage would be a national default of historic proportions in solar system exploration." Van Allen and his colleagues would find themselves repeating such phrases, mantra-like, to the NASA hierarchy for at least another decade. As the Pioneers probed outward for the heliopause, their scientific patrons back home probed for the outer limits of NASA's administrative indulgence.

VERY LONG DISTANCE CALLING

Even without the issue of conflicting mission requirements, the Pio
neer extended mission faced two enormous difficulties that could
not be placated by aggrieved letters to NASA headquarters: communications and power. More specifically, with *Pioneers 10* and *11* moving ever farther from Earth, the task of picking out their signals began
to strain the boundaries of telecommunications technology. And
with both spacecraft continuing to operate well past their original
design limits, the degradation of their radioisotope thermoelectric
generators and the consequent steady decline in power became major considerations.

Communications, of course, had been recognized as a major
concern even in the early hypothetical studies for deep solar system
probes back in the 1960s. Because they must be small, lightweight,
and not power hungry, spacecraft transmitters aren't very powerful:
20 watts is about the limit of their output. Contrast that with your
favorite local radio station, which broadcasts at thousands of watts,
and consider how far that signal carries before you start losing it.
After a certain distance, the curvature of the Earth blocks the signal
(except for certain frequencies such as shortwave that can bounce off
ionospheric layers in the atmosphere to "skip" for greater distances).
Still, without being picked up, amplified, and relayed onward by
other stations (as, for example, cellular phone signals), radio signals
fade over distance and as they fade become more susceptible to interference from other stronger, closer signals on the same or nearby
frequencies.

All these phenomena are compounded over interplanetary distances. The already-weak signals from a spacecraft, such as those from
Pioneer 10's 8-watt transmitter, fade almost into nothing by the time
they cross the hundreds of millions of miles of space to Earth—and
the distance is constantly increasing. The total energy from *Pioneer
10*'s radio signal would have to be collected for several million years
to light a single 7.5-watt nightlight for a millionth of a second. Only

the largest antenna dishes on the planet are able to gather such infini-
tesimal wisps of signal, and only the most advanced amplifiers, sig-
nal-processing equipment, and computers are able to extract and
discriminate that signal amid all the random radio noise of the galaxy. Space may appear dark and tranquil in visible light, but it's a
cacophony of noise in radio frequencies. Using relatively quiet frequencies for spacecraft communications helps, but there's always
some stray noise through which the desired signal must be detected.
And simply picking up a signal isn't enough, not if you want to do
useful scientific work. For that the radio signal must carry data, and
the rate and amount of data that can be transmitted are also dependent on the quality of communications. If the signal is weak or noisy,
only a limited amount of scientific data can be encoded onto it, else
the data would be mostly lost in transmission. The data rate can be
decreased, even to a trickle, to ensure that the data get through, but
doing this means that it takes much longer to send it, which means
longer tracking time on the DSN, time that has to be taken from other
missions.

Improving the technology is one way to compensate. Since its
inception, the DSN has been at the forefront of telecommunications
technology, continuously upgrading its antennas, receivers, and other
equipment and improving their sensitivity. As it happened, Pioneer's
extended mission may have proved to be both an incentive and a
rationale for doing so. "JPL was in the communications business.
They wanted to improve communications technology," Richard
Fimmel explained. "They used maintaining communications with
Pioneer 10 and *11* to justify additional development money from
headquarters to improve the DSN. So Pioneer [made it possible] that
they could do these things, because there was no one else that far out.
We were the biggest challenge to their technology."

Some at JPL might argue that because the Pioneer extended missions were no longer high priority, it was really the requirements of
other, more contemporary missions that drove the DSN upgrades. In
any case, there's no doubt that keeping in touch with *Pioneers 10* and

11 was a great technical challenge for the DSN and one that was made easier by the improvements. In 1982, for example, the DSN began a program to enlarge its three 64-meter antennas to 70 meters in diameter, greatly increasing their sensitivity and signal-gathering capacity. This modification alone extended the useful lives of the truly distant spacecraft such as the Pioneers and Voyagers. As DSN engineer Douglas J. Mudgway wrote in *Uplink-Downlink: A History of the Deep Space Network*: "For most of their long lives, the *Pioneer 10* and *11* spacecraft were able to benefit from the periodic enhancements to DSN downlink capability which were driven not by the Pioneers but by the requirements of other spacecraft. In that sense, they never became obsolete and never reached the predicted limit of the DSN capability to maintain an uplink and a downlink with them."

Part of the Earth-spanning Deep Space Network, this 70-meter dish antenna at Goldstone, California, helped maintain the tenuous link with the Pioneer spacecraft.

While the DSN could beef up its facilities on Earth to maintain its link with Pioneer, there was nothing that could be done to stave off the inevitable decline in the output of the spacecraft RTGs. It wasn't a question of fuel depletion; the plutonium dioxide capsules in the RTGs would continue to generate radiation for many thousands of years. The problem was that the radiation itself steadily degraded the thermocouples inside the RTGs that use the radiation-created heat to generate electricity. As the thermocouple material degraded, the electrical capacity of the RTGs diminished, and with it their power output. And as power levels slowly but surely continued to drop aboard both Pioneer spacecraft, some tough decisions had to be made by the new Pioneer project manager, Richard Fimmel, and his team. Because the spacecraft systems, including the transmitter, needed a certain minimum level of power to keep operating, something else had to be turned off when there was no longer enough power for everything.

Something else had to be some of the science instruments. Some of the instruments had already been silent for some time, since there was little or nothing for them to do anyway. Of the instruments still operating, however, such as Van Allen's Geiger tube telescope and Simpson's charged particle package, the need to ration power meant that they could no longer be turned on all the time. This, in concert with the decreased tracking schedules, meant less data coming in for the active PIs. Although the scientists realized there was nothing to be done about the physical exigencies of diminishing power and increasing distance, these factors, along with the possibility that contact could be lost with the spacecraft at almost any time whether for reasons technical or political, inspired them to redouble their efforts to maintain a reasonable share of DSN time for Pioneer. "John Simpson and I sort of led the charge to keep on getting data on cosmic rays, solar particles, magnetic fields, the interstellar medium," James Van Allen said.

In fact, Van Allen and Simpson were tireless, persistent, and no doubt rather annoying to NASA headquarters in their campaign to

keep NASA from forgetting about Pioneer. Whenever NASA or Congress made noises about pulling the budgetary plug on the Pioneer
extended mission, they were certain to be deluged by protests from a
legion of scientists, many of them not even directly involved with the
Pioneer project but with a keen interest in the data it provided. Van
Allen, Simpson, and their colleagues made some compelling arguments. A February 1977 "Draft Resolution for the NASA Interplanetary Sciences Working Group" by Simpson didn't mince words. "We
strongly urge NASA to return *Pioneer 10* to an average telemetry coverage by the DSN of approximately 8 hours per day. . . . We base our
fear and concern for future coverage for *Pioneer 10* on the 'track
record' of NASA for past coverage of *Pioneer 10*. . . . Although outstanding scientific discoveries have been made in this period of de-
clining telemetry, we are now approaching a disaster level at which
meaningful scientific research becomes impossible."

In a 1981 memo to Fimmel pointedly entitled "Why Is It Vital for
NASA to Track *Pioneer 10* and *Pioneer 11*," Simpson observed that
Pioneer 10 was in a unique position for observations, on a trajectory
"where no other spacecraft is planned to go in the foreseeable future."
It would also be the space probe farthest from the Sun until 1989, and
in conjunction with *Pioneer 11* and the two *Voyagers*, was vital to
forming a large-scale picture of the heliosphere. And Pioneer was not
only scientifically important, it was practical: "Given the severe financial constraints on the space sciences during these years, it is clear
that the continued acquisition of data from *Pioneer 10* offers one of
the most cost-effective ways to maintain a program of discovery and
exploration in the solar system and to assure continuity in the support of investigators in the space sciences."

Van Allen also appealed outside NASA to the scientific and technical community in general as well as the public. "One of the most
incredible features of the fiscal 1983 program of the National Aeronautics and Space Administration is the premature termination of
the deep-space missions of *Pioneer 10* and *Pioneer 11*," he wrote in an

editorial published in *Aviation Week & Space Technology* in April 1982. "The annual saving is $3 million or four ten-thousandths of the agency's budget." Van Allen reiterated Pioneer's many achievements and the scientific payoffs it continued to promise, including the possible discovery of the heliopause and passage of the spacecraft into true interstellar space: "This crossing, if observed, will be the crowning achievement of its extended mission and a milestone in human achievement."

The support of such prestigious scientists, not to mention the relatively low cost of keeping the extended mission going, kept the Pioneer project alive—barely, and with the ever-present threat of the plug being pulled—long enough for *Pioneer 10* to become the first spacecraft to leave the solar system in June 1983. On June 13 it crossed the orbit of Neptune, at that time the farthest planet from the Sun, since Pluto's highly eccentric orbit takes it inside Neptune's orbit for part of its trip around the Sun. To those such as Van Allen searching for the heliopause, of course, *Pioneer 10* was still technically inside the solar system, that is, within the Sun's sphere of influence, but such a definition was much too esoteric for NASA and the public at large, for whom the outer boundary of the solar system meant the farthest planet from the Sun, period. In any case, there was no doubt that it was a significant milestone, one that was duly noted by the press, the public, and even James Beggs, the NASA administrator, who admitted to an audience at the National Air and Space Museum that he had been ready to close down the project until Van Allen and the scientists convinced him otherwise.

If Beggs needed more convincing, *Aviation Week* reported that "there were celebrations of [*Pioneer 10*'s passage out of the solar system] in various cities, with playback of the telemetry that now takes 4 hours to pass from spacecraft to Earth. A special telephone number established for the public to call and hear the same thing was swamped the day it was activated. Planetary space also has its public appeal. . . . Bon voyage to *Pioneer 10*. Bring on the new planetary explorers to follow her." At the Pioneer Mission Operations Center at Ames, the team gathered to bid *Pioneer 10* farewell. "Tomorrow, on to the stars," Jack Dyer said, raising a glass of champagne.

Aside from its main new mission of searching for the heliopause, *Pioneer 10* had been doing some other interesting things. It verified that Jupiter has a huge, fluctuating magnetic "tail" that extends as far as the orbit of Saturn. In tandem with *Pioneer 11* on the other side of the solar system, it searched for gravitational waves predicted by Einstein's General Theory of Relativity, a phenomenon that would probably be very weak and detectable only over vast distances. And perhaps most intriguing, both spacecraft had been investigating the possibility of a tenth planet or other even more exotic object beyond Pluto. For years, astronomers had noticed some odd perturbations in the orbits of Uranus and Neptune, which could be caused by the gravity of another massive object in the outer solar system. Pluto was found to have insufficient mass to affect the orbits of the other planets, so if the perturbations were real, something else was causing them. Speculations ranged from an as-yet-undiscovered tenth planet, to a dark companion star orbiting the Sun, to a black hole.

If something was out there, however, its gravity would also affect one or both of the Pioneer spacecraft. Pioneer PI John Anderson of JPL continued to analyze Doppler data from *Pioneers 10* and *11*, looking for changes in their trajectories and speed that would indicate the gravitational pull of an unknown object. The spin-stabilized configuration of the Pioneers made them especially well suited to such a task because they would always maintain the same course and speed unless maneuvered by commands from Earth—or affected by an outside object.

Meanwhile, one other Pioneer spacecraft had achieved a position of prominence by traveling no farther than 3,000 miles across the United States, from California to Washington, D.C. In January 1977 the third *Pioneer F/G*-type craft prototype built by TRW, used for testing and as a backup, was donated by NASA and the company to the Smithsonian Institution's National Air and Space Museum. It hangs in the Milestones of Flight Hall, in the company of the Wright Brothers' Flyer, Charles Lindbergh's *Spirit of Saint Louis*, and the *Apollo 11* command module. Contrary to popular belief, the Pioneer spacecraft hanging in the Smithsonian is not a model. It is actual flight hardware (with the exception of the RTGs, of course, which are dummy replicas that generate absolutely no troublesome alpha radiation to endanger the tourists). As proud as they might be to see the spacecraft and their instruments on display in the Smithsonian, some of the PIs waged a brief campaign to send the final Pioneer spacecraft on a much longer trip.

THE PIONEERS THAT NEVER WERE

Maybe it was the predilection of the various Pioneer spacecraft to continually achieve more than originally intended. Maybe it was their reliability and economy. Maybe it was because the Pioneer team knew how to get things done. Or maybe it was just coincidence. Whatever the reasons, the Pioneer project seemed to inspire quite a few ideas for spinoffs, either extensions or modifications of existing missions or new missions that used Pioneer as a foundation and an inspiration.

One major example is the modification of the *Pioneer 11* mission by sending the craft on to Saturn, certainly not part of the original mission plan but an option that became attractive when *Pioneer 10*'s Jupiter success freed up *Pioneer 11* from repeating the same mission. Since the primary objective of the Jupiter encounter had been achieved, there was nothing to lose in attempting to stretch *Pioneer 11*'s mandate a little by giving it one more stop on its way out of the solar system.

Once that decision had been made, controversy arose over the particular trajectory *Pioneer 11* should take to Saturn, as we've seen. But along with the "inside option" that would have taken the craft between Saturn and its rings, and the "outside option" that was eventually chosen for the sake of Voyager, Pioneer navigation expert Jack

Dyer realized that there was one more possibility. Just as Voyager would later use Saturn to boost itself onward to Uranus, *Pioneer 11* could do the same thing. *Pioneer 11* already had two planets in its hip pocket. Should it try for one more, becoming the first spacecraft to visit Uranus and to encounter three planets in a single mission?

The prospect was seriously, if briefly, considered. Whether because of a reluctance to push their luck or to try convincing NASA headquarters and inciting conflict with JPL, though, few in the Pioneer project were optimistic. A 1975 Ames study on the *Pioneer 11* Uranus option was positively glum. "December 1985 is earliest possible arrival [at Uranus]. More durable life than any known previous spacecraft is needed," it stated. "Productive potential at Uranus may be academic due to the low probability of delivery there of a surviving spacecraft with sufficient electrical power and strong enough radio signal to be useful." Van Allen remembered: "I was attracted by this possibility, and [Jack Dyer] and I discussed it at length but quietly abandoned it on two grounds." The first was that the trajectory required to send *Pioneer 11* to Uranus would have compromised the scientific observations to be made at Saturn. Second, Van Allen recalled, "There was a significant probability that the spacecraft would not survive either the Saturn encounter or the long additional flight time to Uranus." In hindsight, of course, such fears were much too pessimistic, but they seemed perfectly reasonable at the time. The Pioneer team decided to settle for one unplanned bonus and wring everything possible from it, rather than gamble on such a long shot.

Still, the possibility is tantalizing and could be considered a classic example of a missed opportunity. If it had been attempted and been successful, such a "mini-Grand Tour" of Jupiter, Saturn, and Uranus by *Pioneer 11* would have been a most spectacular achievement, at least technologically if not scientifically. But the scientific return at Uranus would probably have been minimal, hardly comparable to that from *Voyager 2*'s 1986 encounter, and *Pioneer 11*'s Saturn encounter would have been more distant and thus less memorable itself. However tempting the idea of the Uranus option may have been at the time, the decision to abandon it was probably the best one.

Even before *Pioneers 10* and *11* left Earth, there were some who argued that sending merely two spacecraft wasn't enough, particularly when there was a perfectly good third spacecraft waiting in the wings. Along with the two designated flight spacecraft, *Pioneer F* and *Pioneer G*, TRW had also constructed a prototype craft used for engineering tests and other measurements and had spare parts for yet another spacecraft. Instead of allowing this spacecraft to sit idle in Redondo Beach, why not launch it on another mission that would complement the other Pioneer flights, out of the ecliptic plane to make the first dedicated investigation of the heliosphere and the solar poles?

Scientifically speaking, it was a wonderful idea. No probes had yet ventured high above the ecliptic plane to examine the nature of the solar wind at such latitudes or to take a direct look at the polar regions of the Sun. Such a spacecraft would supplement the observations of *Pioneers 10* and *11* as they probed outward into the heliosphere, as well as the flow of data from *Pioneers 6* through *9*. It would provide a much broader and more complete picture of the total solar system environment than had ever been possible. Best of all, the out-of-ecliptic mission would allow one more close flyby of Jupiter because the probe would use a Jovian gravity boost to hurl itself out of the solar system's plane.

A 1971 Ames study also spelled out the technical bonuses. "First, *Pioneer F/G* technology, still current, is especially appropriate for this mission and mission requirements appear to be well within the capability of present Pioneer design. Second, by using the Pioneer prototype spacecraft, which was designed and tested as a flyable unit, a relatively low mission cost can be achieved. Third, since it is a continuation of an ongoing program, it can be phased into the activities of the existing Pioneer Project office without additional personnel requirements." The study projected a 1974 launch with essentially the same experiment payload as the *Pioneer F/G* spacecraft; the out-of-ecliptic mission would have been designated *Pioneer H*, becoming *Pioneer 12* after launch.

The preparations for the already-approved *Pioneer F/G* missions placed consideration of the proposed *Pioneer H* on the back burner and ended any thoughts of a 1974 launch, but the PIs persisted in pushing the mission. "We mounted an intensive campaign to launch the flight-worthy spare spacecraft and its instrument complement on a low-cost, out-of-ecliptic mission via a high-inclination flyby of Jupiter," Van Allen recalled. "However, our case fell on deaf ears at NASA headquarters, and the spare spacecraft now hangs in the main gallery of the National Air and Space Museum, at 1 AU [astronomical unit, equivalent to the distance of Earth from the Sun or 93 million miles] and zero ecliptic latitude." The explorers of the heliosphere such as Van Allen would have to wait considerably longer for their dedicated out-of-ecliptic mission, finally launched as *Ulysses* in 1990 in a joint NASA-European Space Agency project.

After the first successful Jupiter encounters by Pioneer, scientists and engineers began to consider the next step: not simply flying past the planet and taking a quick look but orbiting it for a more detailed survey. And more than that, an orbiter could be equipped with a separate probe that could be sent into Jupiter's atmosphere. The idea became known as the Jupiter Orbiter-Probe or JOP, and Ames was one of its earliest champions, seeing the mission as a natural task for Pioneer's proven technology. "Both the probe and its scientific instrumentation are contained within the inherited experience from the Pioneer Venus program," said a 1975 Ames study. The illustrations of the proposed JOP spacecraft, in fact, look like an uncanny hybrid of the *Pioneer F/G* design with an attached *Pioneer Venus* probe, an impression the study confirms: "The PJO/P spacecraft is a derivative of the *Pioneer 10/11* design. . . . Residual hardware from the *Pioneer 10/11* program, with some additions, can be used in an engineering/prototype spacecraft for early development tests."

During the next decade, the JOP mission would suffer through one of the most tortured development histories ever seen in NASA, a victim of partisan politics and budget wars and several near-cancellations, before it evolved into the *Galileo* mission. Then more political wrangling, equipment modifications, and plain red tape, some caused by the 1986 *Challenger* disaster, delayed its launch until 1989. Yet despite all its problems, *Galileo* went on to remarkable success, its mission exceeding all expectations until it was finally ended in 2003. And although the *Galileo* orbiter was a JPL project and not ultimately derived from the Pioneer design, the atmospheric probe was a direct outgrowth of Ames and Pioneer work on probes and heat shield technology, including the extensive experience with *Pioneer Venus*.

The *Galileo* probe project was managed by Ames, originally under Joel Sperans, one of Charlie Hall's proteges. "He was my science manager on *Pioneer Venus*," Hall said. "It was on-the-job training for him. Most of the things he thought were good on *Pioneer 10* and *11*, he used on *Galileo*. A lot of people have that 'not invented here' attitude, but he didn't have that. I went out and looked at some specifications he was writing for *Galileo*, and you could have changed the name from *Galileo* to *Pioneer*, and it would have looked the same. The whole format and approach were what we had started." The influence of Charlie Hall and the Pioneer project was still in force, years after both were officially put out to pasture.

The Pioneer spacecraft themselves continued to operate, however, and the PIs continued to sift through and analyze data from the planetary encounters and beyond. For at least one scientist, the Pioneer data led to a controversial new theory that, if correct, could upend the entire science of astronomy.

SOMETHING IN THE DARK

Robert Soberman wasn't one of the more prominent Pioneer PIs, and he was well aware of that fact. His instrument, the Sisyphus asteroid-meteoroid detector, had hardly been involved in the high-visibility planetary encounters, had been essentially put out of commission by the intense radiation of Jupiter on both spacecraft, and had provided some rather odd data that didn't correlate well with the findings of the other onboard experiments that observed meteoritic particles, Kinard's pressure cells and Gehrels's IPP. "The Sisyphus event data that came back were nothing like what we expected," Soberman said.

In fact, they differed from the data provided by another mode of the Sisyphus instrument by a factor of 10, an inexplicable discrepancy. Soberman's data were attributed to instrument error or signal noise, and the subsequent papers he published were mostly discounted by his colleagues. "There was a fellow by the name of Siegfried Auer, who I must have slighted in some way or another. He made it a vendetta to attack the Sisyphus data, and quite successfully," Soberman remembered. Auer's papers disputing Soberman's results appeared in *Science* and the *Journal of Geophysical Research,* prestigious journals with high visibility. As if to make his objections unambiguous, Auer titled his *Science* paper "Doubts About the Particle Concentration Measured with Asteroid/Meteoroid Detector on *Pioneer 10.*" Apart from Auer's arguments, the failure of his asteroid-meteoroid detector after Jupiter (even though expected), not to mention the experiment's inadvertent interference with the other science instruments, also didn't help Soberman's cause in the scientific community. Although his team's proposal for another meteoroid experiment on the Voyager missions made the first cut, it was ultimately rejected. Neither Voyager spacecraft carried a meteoroid experiment—unfortunately for Soberman, who said that another instrument "would likely have validated the Sisyphus data."

Soberman himself was puzzled by his Sisyphus results, although he was certain that his data were valid and not simply a technical anomaly of his instrument. "That's the standard scientific way of answering anything you don't understand—it's system noise, an artifact of the system," he said. A colleague at NASA, Maury Dubin, helped Soberman obtain NASA funding for a more detailed study of the Sisyphus results. That led both men to an unexpected conclusion. "Maury convinced me that Sisyphus has measured a new population of meteoroids that came from outside the solar system, interstellar meteoroids," Soberman said. They called these objects "cosmoids," a contraction of cosmic meteoroids. "In 1991 we published a paper saying once you throw out the idea that you're measuring interplanetary material and recognize that you're being dominated by this population, all three [Pioneer meteoroid] experiments agree, and agree down to fine details."

Soberman realized that this idea of interstellar meteoroids or cosmoids had implications more profound than might seem obvious at first glance. "We are contending that Pioneer discovered the dark matter of the universe, and that it's nothing strange or exotic. It's just these dark interstellar cosmoids. And once you recognize their existence, a lot of things follow." In fact, Soberman argues, "Once you recognize that this population exists, just about everything that is taught in astronomy is wrong. In some cases, 180 degrees out of phase with nature."

Soberman and Dubin have no illusions about challenging the scientific establishment: "We might be wrong. We might be crazy," Soberman agrees. But he also insists that if their conclusions are correct, much of the foundation of modern astrophysics will have to be revised. His cosmoid dark matter theory challenges astronomical tenets as fundamental as the process of stellar formation, the Big Bang, the expansion of the universe, and the red shift.

Such sweeping claims are sure to arouse the ire of other scientists, and profound skepticism is indeed the proper response to a theory so much at odds with established scientific knowledge. Soberman acknowledges and welcomes such skepticism, fully knowing it's all part of the process. The strength of science is that it's never absolutely set in stone; it changes and evolves with new discoveries, but slowly and gradually, only after the new is proven to be valid and either integrated with or supplants the old. Unless and until his cosmoid theories are verified and accepted by the general scientific community, Soberman continues to teach the standard and accepted astronomy texts in his university courses.

Soberman is unlikely to be proven correct anytime soon. The cosmoid theory challenges so many basic models that an overwhelming preponderance of evidence in its favor would have to be accumulated by many scientists before it could finally be accepted. It may end up as no more than one of myriad interesting hypotheses destined to be footnotes in the history of science. If even some of its assertions are eventually proven valid, however, then *Pioneer 10* may be responsible for a discovery more profound than any of its other accomplishments. If correct, the cosmoid theory would be the ultimate example of Pioneer far exceeding expectations yet again.

CLOSING THE BOOKS

After all the intimations of danger, the fears of being pulverized by an asteroid, shredded by dust particles, or fried by radiation, the ending came quietly, with a wistful tranquility. In late 1995, after almost 22 years of life and contact with home, *Pioneer 11* disappeared into the darkness.

Although its RTGs seemed to have been deteriorating at a somewhat faster rate than *Pioneer 10*'s, it wasn't a lack of power that was responsible for *Pioneer 11*'s loss of contact with home. The fatal problem was the failure of the spacecraft's Sun sensor, which was essential to provide an orientation reference to keep the spacecraft's main antenna pointed toward Earth. Because the Sun, as seen from *Pioneer 11*, slowly drifted across the sky as the spacecraft continued speeding away, and because the Earth's position changed as it orbited the Sun, it was necessary to periodically reorient the ever-spinning spacecraft by firing its maneuvering thrusters to precess its spin axis and keep the antenna pointing toward Earth. Otherwise, the already faint signal from *Pioneer 11* would fade into nothingness. But the maneuver required an operating Sun sensor so that the repointing maneuver could be accurately planned. Without the ability to keep the spacecraft pointed toward Earth, contact diminished until its signals were beyond even the detection capabilities of the DSN. In the end, *Pioneer 11* didn't go out in a blaze of glory on a suicide mission inside Saturn's rings; it just faded away, heading toward the constellation of Aquila the Eagle, having forever lost sight of home.

Pioneer 10 was still calling home and still sending back data. But the ever-increasing distance was taking its toll, both on the amount of data received and the willingness of NASA headquarters to keep listening. Fred Wirth, who had become project manager in 1995 upon Richard Fimmel's retirement, explained: "Unfortunately, with the spacecraft distance, the bit rate, the transmission rate, went down. We started out at 2,048 bits per second, and by that time we were down to 8 or 16 bits per second of real-time data rate. . . . Now when you imagine the accumulated number of hours that we tracked and you translate that into 8 bits per second from one instrument, the scientific value of the spacecraft wasn't worth it."

Although Van Allen and Simpson, whose instruments were the only two still in operation on the spacecraft (and still searching for the heliopause), continued to fight for *Pioneer 10*'s survival, the project had reached and passed the point of diminishing returns, at least as far as the NASA brass were concerned. "I had nothing but budget troubles with headquarters," Wirth remembered. "They just wouldn't give me any more money. I'd have a 10 million dollar a year budget, and with that I'd finance all the PIs, all their data analysis, computer time, grad students. So a lot of money was going out to the PIs. If you look at the costs of the mission control staff, it was still a lot of money. I think I was finally down to about 2 million dollars a year and a reduced staff, and NASA headquarters said, you can't do that, we just don't have the budget, and they threatened to cancel the program. I fought it and got it extended and got more money for about 2 years. From '95 to '97 we really fought for every buck."

And, as always, time on the DSN. "I used to go down there every three weeks negotiating tracking schedules," says Wirth. "I would have to beg and ask and cajole with other project managers of JPL missions. They all thought their project was more important than Pioneer. I had to fight for every hour of tracking we could get. I had to do some tough negotiating. But I got my reasonable tracking."

In 1997 the patience of NASA headquarters ran out. NASA officially announced that the Pioneer project would end on March 31, just after the celebration of the craft's 25th anniversary. Wesley Huntress, NASA associate administrator for space science, proclaimed that "*Pioneer 10* exemplifies the American pioneering spirit of exploration far beyond the frontier. Not only has it made many major scientific discoveries in the far reaches of space, we're proud that it has managed to stay alive almost 10 times longer than the original mission called for, a tribute to the designers and builders at TRW and the operators at NASA's Ames Research Center." He added: "NASA operated the *Pioneer 10* mission as long as it had enough power to return science data about the conditions in space as far from Earth as possible. We will end the science mission at the end of March because the power has finally become too weak to do significant science."

On March 3, special events were held simultaneously at Ames and at NASA headquarters to honor the achievements of *Pioneer 10* and the Pioneer program in general. The conference at headquarters provided a stunning example of just how far technology had advanced since *Pioneer 10*'s launch. It was attended not only by an in-person audience but also by a huge virtual audience of at least 10,300 on the Internet, who could watch the proceedings on home computers and ask questions of the conference participants. Later in the evening, Hall and Van Allen spoke at a reception at the National Air and Space Museum.

For the Pioneer veterans who spoke at the Washington conference, including Charlie Hall and many members of his team, along with several of the PIs, it was both a proud and bittersweet affair. Some still weren't quite ready to say goodbye, including James Van Allen, whose instrument was the last one operating. "I was one of those who gave a eulogy on the achievements of Pioneer, and meanwhile I was working the hallways trying to keep it going," he recalled.

Van Allen's efforts were for naught, however. Reporters were invited to Ames on March 31, 1997, to watch the final signals being received from *Pioneer 10*, along with the last bits of science data from Van Allen's Geiger tube telescope. The DSN station in Madrid did the honors, beginning its track of the spacecraft at 5:25 A.M., California time. At precisely 11:45 A.M., controllers watched the signal strength indications of the *Pioneer 10* downlink fade and drop off their screens. In an elegiac bit of symbolism, the lights in the Pioneer Mission Operations Center were turned off at that moment to mark the occasion.

It was all over. Officially.

But unofficially, not quite yet.

12

Lone Survivor

I t was still out there and still alive, official pronouncements of its death notwithstanding. The difference was that now it was going to be even more difficult to talk to *Pioneer 10*, and not just because of administrative myopia. Larry Lasher, the man to whom Fred Wirth had passed the Pioneer torch upon retirement, faced some fundamental hard truths of physics, technology, and politics if he was to hang on to the spacecraft's tenuous link with Earth.

It helped that *Pioneer 10* had some allies, if not so much in NASA headquarters anymore, then in other places of more immediate importance to the challenge at hand. "We had a lot of connections and friends at JPL," said Fred Wirth. "They managed to squeeze in an hour or so per week of tracking coverage and maneuver whenever we could." It would have to be enough, because just as in Pioneer's official "heliospheric mission" phase, the greater the distance, the weaker the signals and the harder for the Deep Space Network to lock on. Van Allen was still getting data occasionally but at a painfully slow rate. The only reason his Geiger tube telescope was still in operation was, as he explained with a touch of pride, "[it] was the most parsi-

monious instrument as far as power requirements, the most parsi-
monious on the spacecraft, actually."

But in every respect, *Pioneer 10*'s link to home was a thin, steadily
unraveling thread. If Lasher was to prevent it from breaking, he was
going to need a lot of resourcefulness and a little luck. Fortunately, he
had both.

DIEHARDS

Appropriately enough, Lasher's first bit of luck came to him at least
partly because of Pioneer's own legacy. In 1998 the record of accomplishment and ability to design and manage space missions demonstrated by the Pioneer project brought the *Lunar Prospector* mission
to Ames. One of Dan Goldin's new low-cost "Discovery" projects,
Lunar Prospector was intended to orbit the Moon's poles and map its
surface composition and gravity, along with looking for signs of water ice that had been spotted by the earlier Clementine spacecraft.

Shrewdly, Lasher found a way both to further the *Lunar Prospector* mission for Ames and to give *Pioneer 10* a new lease on life. He
convinced NASA that tracking *Pioneer 10* would be a perfect training
exercise for *Lunar Prospector*'s controllers. For a couple of years,
Pioneer's old control room hummed again with data from *Lunar
Prospector* and *Pioneer 10*, until the *Prospector* project ended when
the craft was deliberately crashed into the Moon in July 1999, in hopes
that the impact would kick up water vapor that could be detected by
spectroscopic analysis (it didn't).

Prospects for *Pioneer 10* again looked bleak, but Lasher and the
other Pioneer diehards were used to that by now. And again, they got
lucky. "A white knight came along in the form of a NASA headquarters study on weak signals," Lasher explained. With more deep-space
missions in the pipeline, NASA was trying to develop new techniques
of extracting useful telemetry from very faint signals using computer
algorithms based on chaos theory. The study found *Pioneer 10*'s faint

chirp a perfect subject, providing a few more dollars of funding from headquarters for a short while.

Meanwhile, another angel of sorts helped to keep the candle burning for *Pioneer 10*. For a few years the SETI Institute in Mountain View, California, tuned into *Pioneer 10* from the giant radio telescope at Arecibo, using the probe's signal as a fair simulation of a hypothetical communication from an extraterrestrial civilization and to calibrate its SETI searches. In a 2000 interview, James Van Allen noted: "One of my colleagues, an astronomer, visited Arecibo about a month ago, and he brought back a sample of the recorded signal from *Pioneer 10*, which is loud and clear. They're using it for calibration of the detectors they use for SETI. It's a very valuable calibration source for them, because it has a known power output and is used to determine the sensitivity of their instruments at that particular frequency." Although Van Allen said that as far as SETI was concerned, "I'm not holding my breath for affirmative results," he contended that it was "another incidental virtue of keeping *Pioneer 10* going."

But despite the old axiom, while there's a will there's not always a way—at least, not forever. While the will of scientists, Pioneer project veterans, and SETI researchers persisted, the way was rapidly eroding. More than the steadily weakening signal, which would have been lost years earlier without the various upgrades to the DSN's antennas, receivers, and other equipment. As its 30th anniversary approached, *Pioneer 10* found itself becoming rapidly overtaken by the same phenomenon that had revolutionized nearly all forms of technology, for better or worse: the digital age.

Pioneer 10 had been conceived, designed, and constructed in a world when computers filled air-conditioned rooms, when interfacing meant punchcards and magnetic tape, when the only mouse ever seen in a computer center was of the cheese-eating variety. Because it carried no actual computer onboard, only some registers that could store and execute a limited number of commands, all of the computing of Pioneer, including the processing of data and the compiling and execution of spacecraft commands, was done by computers back

The Pioneer Mission Operations Center at Ames.

on Earth, linked to the spacecraft by nothing more than the DSN uplink and downlink. As those computers began to shrink in size and pick up in processing speed in the years since *Pioneer 10* left Earth, their ability to talk to the spacecraft began to be affected.

By 2000 the project's remaining original DEC PDP 11-14 computer, standing like a museum piece in the Pioneer control room at Ames, was the only operating computer that could be used to process and transmit navigational commands to *Pioneer 10*. Operations supervisor Ric Campo lamented, "The equipment is barely keeping alive. The tracking station equipment that is used for *Pioneer 10* is also obsolete. The equipment and data interface are outdated, and there are little to no monies to maintain it. Much of the hardware is maintained by stripping cards and interfaces from similar hardware in the Pioneer control center." Although the data telemetry system was reconfigured several years ago to run on a Macintosh rather than the old mainframe, such a fix isn't possible for the command system. Present-day engineers can still interpret the data sent by Pioneer's ancient systems, but the reverse isn't true: Pioneer can't understand commands not compiled by the computers of its own age.

Aside from the obsolescence of its hardware, *Pioneer 10* also faced the inevitable bugaboo of all computer users: software upgrades. Lasher explained: "We have a specific software arrangement that the DSN uses to contact our spacecraft, and they're changing it over in 2001 or 2002. Even if we wanted to, we couldn't be supported because the DSN software wouldn't be able to track it any longer."

Finally, there remained the one problem that literally grew worse every second, as the distance between *Pioneer 10* and Earth grew. The signal was always fading, fading, fading, inexorably approaching the absolute limit of the DSN's ability to detect it. And as the distance grew, so did the two-way communications time. By 1999, with *Pioneer 10* over 6 billion miles from home, that interval was over 20 hours. When controllers sent a command to the spacecraft, they had to wait that long to ensure that it had been properly executed.

The wait itself wasn't a problem. Pioneer's controllers had nothing but patience. The eternal sticking point was DSN scheduling, particularly for a project that was low priority and no longer had any official status. Getting time for one 70-meter dish track on the DSN was hard enough for Pioneer. But also getting time for a second track exactly 20 hours later—a track that had to be at a different DSN station because the rotation of the Earth would have carried the first one out of sight of the spacecraft in the interval—was a trick that often took some cajoling, politicking, and good old-fashioned favor trading. All too often, such efforts didn't work. "DSN has been gracious enough to keep providing us some support, but there are too many other high-priority missions," Lasher said.

Yet somehow, as the 21st century dawned, Lasher, Campo, Dave Lozier, and a few other Pioneer veterans managed to keep it going. "No one seems to want to let it go, even if it requires coming in after hours or on weekends," said Lasher. "Some will come in the middle of the night because that was the time commands had to be sent up." Speculation as to how long contact with *Pioneer 10* would last became a favorite topic of conversation around the old Pioneer Mission Operations Center at Ames, although few were willing to make any bets. Every time the end was darkly pronounced by a NASA official, engineer, controller, or scientist, the deadline came and went, and the next DSN track of *Pioneer 10* still managed to pick out the spacecraft's voice amid the galactic radio noise. The doomsayers were like some apocalyptic cult, periodically assembling on a remote mountaintop after shrilly proclaiming the imminent end of the world, only to have the obstinate Sun rise anyway the next morning.

Pioneer 10's survival for 10 years had been extraordinary. Its continued life after 15, 20, 25 years was nearly miraculous. But now it was approaching the 30-year anniversary of its departure from Earth. Signs of the end were beginning to show.

By the summer of 2000, even some of the true believers started to think that the thread had snapped at last. In August the periodic antenna repointing command sequence was sent to *Pioneer 10*. But the DSN had been unable to achieve a two-way lock onto the spacecraft for verification of the maneuver. No one knew whether the maneuver had succeeded, whether *Pioneer 10* had forever lost its lock on Earth as had *Pioneer 11*, or whether the spacecraft was simply dead. Several more attempts to pick up *Pioneer 10*'s carrier signal over the following months failed.

No one was quite willing to declare it all over yet, though. Finally, toward the end of April 2001, controllers tried transmitting to *Pioneer 10*. This time they received an answer, demonstrating not only that the spacecraft was still healthy but also the reason for its silence.

"The one-way oscillator aboard the spacecraft had failed," Lozier explained. "When the spacecraft's not receiving a signal, it'll transmit over an auxiliary oscillator, a crystal device at a certain frequency. That frequency will change with temperature, so it's not a very stable frequency. When the DSN tries to find it they have to do a little search, because the spacecraft continues to get colder. Of course, it's so cold now, it's a wonder the oscillator has worked as long as it did. It has a little heater that supposed to maintain it at a certain range."

Unfortunately, without *Pioneer 10*'s one-way oscillator, the only way for Earth to hear from the spacecraft was to talk to it first. "When the spacecraft is receiving a signal from the DSN, it'll use that uplink frequency that it receives and convert it to a downlink transmission," Lozier said. "So whenever the DSN transmits, they know what frequency, and of course it's Doppler compensated, and they can tell precisely where to look for the signal on the downlink. That's what they call coherent mode of transmission, two-way, up and down. The DSN can lock right on the signal, so they know right where it is."

But to accomplish this, two "tracks" have to be scheduled on the DSN—one to transmit a signal and another to listen for *Pioneer 10*'s response at the round-trip light-speed time of almost 24 hours later. "That's the only way we'll be able to communicate with it," Lozier said. "It takes two tracks. You have to transmit *and* receive. And to get two tracks separated exactly by the round-trip light time is next to impossible given the overloading of the 70-meter network."

Those two-way tracks were indeed few and far between throughout 2001. There was only one more, on July 9. But as the 30th anniversary of *Pioneer 10*'s launch approached, even NASA headquarters didn't argue with making another attempt at contact to commemorate what was inarguably one of NASA's finest achievements. On March 2, 2002, the DSN station at Madrid signaled *Pioneer 10*, by then almost 7.5 billion miles away. Twenty-two hours later the response arrived, right on schedule. Data were received from Van Allen's Geiger tube telescope, dutifully recorded, and sent on to him at the University of Iowa. *Pioneer 10* was still on the job 30 years after launch. "As an eternal optimist, I was confident it would succeed," said Lasher in a NASA press release. "*Pioneer 10* has been discounted in the past, but somehow it always manages to land on its feet."

But as pleased as they were that *Pioneer 10* had made it to its 30th birthday, Lasher, Lozier, and the rest of the Pioneer diehards knew that it was only a matter of time and that the last act of the Pioneer story was nearly over. It was only a question of when the curtain would fall.

LEGACIES

How does a spacecraft built to function for no more than a 21-month mission last 30 years? Is it exceptional engineering, ingenious design, just plain dumb luck? Few who worked on Pioneer will discount the value of luck, having been its beneficiaries on many critical occasions during the missions. But they're also quick to emphasize that the successes of the Pioneer spacecraft are due to much more than simple random chance.

When *Pioneer 10* left the solar system in 1983, John Simpson pointed out that "the electronics on *Pioneer 10* didn't survive just by luck. Many years of thought and care went into what was in those days the early phases of chip technology. You will find on Pioneer the most classic examples of chip technology that now have found their way into the industrial world." Sixteen years later, he said, "we were in this mode of faster, better, cheaper" long before it became NASA's guiding mantra in the 1990s. James Van Allen agreed: "As far as I'm concerned, NASA just rediscovered the principle." TRW's Bernard O'Brien was slightly dubious about the "faster, better, cheaper" notion. "My personal opinion is that you can do any two of those and not get into too much trouble. Doing all three is a real challenge." Despite "a few cost problems" on Pioneer, he said, "to have done what we did significantly cheaper would have been almost impossible." Yet at about 100 million dollars total cost, Pioneer was hardly the billion-dollar extravaganza of the more sophisticated missions that followed.

"They were dirt cheap," Fred Wirth observed. "Compared to JPL's Voyager missions or any JPL missions, Pioneers were the Volkswagens of their day."

Pioneer's veterans consistently stress the probe's straightforward design. The spin-stabilized configuration, use of spaceflight-proven systems, redundancy of vital components, and the navigation and communications techniques that allowed ground-based piloting all reduced the risks of irreparable onboard breakdowns or malfunctions. "He didn't want anything fancy, and he didn't want anything untried," Wirth said of Charlie Hall. "All the Pioneers were made spin-stabilized. He deliberately made them simple that way. And not only did that cost a hell of a lot less, but it made them reliable. Look how long *Pioneer 10* lasted and how long *Pioneer 6* lasted. Once you spin it up, it stays. There were no computer chips onboard, so nothing much could fail. There's nothing simpler than a NAND gate and NOR gate and integrated circuit, and you build a register, like a shift register, out of a bunch of AND gates. And that is reliable."

The Pioneer family of spacecraft. From left to right, *Pioneers 6* through *9*, *Pioneers 10* and *11*, the *Pioneer Venus* orbiter, and the *Pioneer Venus* multiprobe bus.

The same philosophy pervaded the design of the science instruments. "In those days we used discrete components, individual transistors, resistors, capacitors, as compared with modern printed circuitry. By present standards fairly primitive instruments, but by the standard of the late 1960s they were up there with the best of 'em," James Van Allen said. Hall explained that aside from the RTGs and some of the science instruments, "the rest of the spacecraft was pretty much standard-type electronics."

O'Brien also cites the project's emphasis on reliability, which involved exhaustive testing of parts and systems before Pioneer ever left the ground. At TRW, he said, "We spent a lot of time, money, and effort on making damn sure those parts were designed, built, and tested before we ever brought them into our house from the vendors." Fred Wirth agreed: "Every single little resistor, every single part, had its whole history. For each tiny 1-meg resistor there was a great big bunch of paperwork and testing. Because you don't want failure in a lousy little 10 cent Radio Shack part to bring down your spacecraft. All of the spacecraft components were fully tested and certified. That made the electronics more expensive but also increased the reliability." Jack Dyer pointed out that "Charlie insisted that wherever there was an automatic system, we'd be able to bypass it. The longevity of *Pioneer 10* was directly enabled by that insistence."

But simple, reliable design is only part of the picture. No matter how impressively engineered, *Pioneer 10* could never have done its job without exceptional project personnel under outstanding leadership. Talk to any of the Pioneer team members, and it won't be long before Charlie Hall's name is mentioned and always with an uncommon fondness, respect, and even awe. "Charlie Hall was really a star," said Van Allen. "He managed the whole thing with a firm hand but was very constructive and receptive to all the crazy requirements we tried to meet. He was outstanding, no doubt about it." Such sentiments are universal among the Pioneer principal investigators. "He was the key to the success of the whole mission," Tom Gehrels said of Hall. "I've seen quite a few other managers and programs that would be bitterly fighting, and Charlie Hall would have none of that. He had the total confidence of all of us because he was so good himself."

Hall's successor, Richard Fimmel, observed another key factor in Pioneer's success. "Charlie was able to motivate people to work hard. Pioneer was blessed in general with people who liked what they did, believed in what they were doing, and were enthusiastic about what they were doing." Hall nearly echoed Fimmel's words: "Pioneer was blessed by so many good people working on it, from the scientists to the lowest technician."

Many of those people still consider Pioneer to be a professional and even personal highlight of their lives. For Tom Gehrels, "It was so nice that I decided to never write a proposal again to go into space. Just like I went on a row boat through the Colorado River and would never do it again. Just like I got married to the one woman I'm still married to and would never do it again. And that is not a joke. If you have an ultimate experience, you wouldn't want to repeat it. I've said to many people that I would not do another proposal for a space mission because it would be like rowing the Colorado River or getting married again."

Fred Wirth has similar feelings looking back on Pioneer. "Sometimes I get awed. You know, when you work every day on a project, everything sort of becomes routine. But then when you talk to reporters and really stand back and look at the big picture, you say, 'How the hell?' We were the first ever to go past the asteroid belt, the first one to reach Jupiter, and now we're the farthest spacecraft out. Then you feel small, and you feel great pride in having contributed to that."

Perhaps awe and pride are the most appropriate responses to Pioneer: awe at the new knowledge it brought us, and pride for the success and sheer audacity of the accomplishment. And maybe the best and most succinct explanation for the success of Pioneer can be summed up by a statement Charlie Hall made in 1999: "It's a lot of incentive to work on something that man has never done before."

THE TAO OF CHARLIE HALL

Without a doubt, Charles Frederick Hall was the focal point, the center, of the Pioneer universe. Pioneer would probably have succeeded without him but not as quickly, not as cheaply, not as reliably. Examples of other space projects before, during, and after Pioneer show how easy it is for such a complex endeavor to drift off track and become embroiled in out-of-control budgets, lack of a coherent direction, endless modifications of hardware and mission objectives, and political wrangling among contractors, scientists, and administrators. A strong and focused project manager may not be able to overcome every hurdle with the grace and aplomb of a Charlie Hall, but he or she is indispensable in keeping it from becoming a technical, administrative, and financial nightmare.

A *New York Times* article published during the *Pioneer 10* Jupiter encounter spelled out the scope of Hall's task in administering the Pioneer project. "At the peak of the project, he was coordinating the work of some 1,300 people, 950 workers at TRW, the 11 teams of scientists, projectory [sic] analysts at the Jet Propulsion Laboratory, 80 Bendix Corporation data processors and control room operators, some 50 Atomic Energy Commission engineers building the spacecraft's nuclear power plant and another 50 engineers and scientists on his own staff." How did Charlie Hall manage it?

While the *Times* profile noted that Hall's "relaxed, old-shoe manner . . . belies an inner toughness" and observed that during the Jupiter flyby "he took time to preside over all news briefings, chat with scientists in the corridors, have a few leisurely drinks with reporters, and take home movies of activities around the control room as the climax of the mission drew nigh," it missed the true core of his managerial philosophy. But those who worked closely with Hall don't hesitate to cite his most important credo because they breathed it and lived it every day on Pioneer: *keep it simple.* "If it's complex, keep it on the ground," Richard Fimmel said. "Keep the spacecraft as simple as possible. Do everything from the ground that you possibly can, which meant we had a lot of hands-on to do. But it doesn't take a genius to figure out that the more complexity you have on the spacecraft, or any instrument or vehicle, the more problems you get. Keep it on the ground if it doesn't have to be on the spacecraft."

Al Eggers makes the same point: "Charlie was very careful not to try and do too many things on a specific spacecraft. Do a few things and really do it right. And this is not only true of spacecraft but most everything. If you're getting yourself out on a limb in terms of risks, be real careful how much you overburden that limb you put yourself out on."

Although Hall certainly knew how to delegate, he also kept a close personal eye on every aspect of the project. "He insisted on seeing every plan and reading it word for word himself, and then calling you into his office to answer every challenge he had to what you wrote," Jack Dyer recalled.

Sometimes, Hall could even be a bit too much of a hands-on guy. Fred Wirth recalled one example from the early days of the Pioneer program. "When I built the control center we produced engineering printouts. So daily I had to deliver to Charlie Hall's office three or four 12-inch binders of computer printouts, a huge stack of engineering data from all night long. There were these humongous binders that we produced of folded computer paper. So every morning we delivered stacks of those binders . . . and I thought that was kind of silly. I kept saying, you know, Charlie, if you want samples of engineering data taken every hour for anything unusual, just ask me for it and I'll get it to you in a few minutes. But no, he insisted, daily delivery. Well, I thought it was ridiculous [that] he wouldn't accept any other solution."

Wirth knew that Hall's stubbornness usually had a good reason but felt that in this particular instance he was being rather extreme. "So finally I had a brilliant idea. I told my guys, 'Hey, don't make any printouts for the next few days.' And I told Charlie, 'I'm so sorry, but the printer's crapped out. We don't have another printer, and we have to send for a spare part.' I made up this elaborate story about why we couldn't print the data. And there was nothing wrong with the printer. So he didn't get any data for 3 or 4 days. Then finally I said, okay, go. And they produced about 30 volumes of binders with printouts, loaded them on to a great big cart and wheeled it down the hall, and said, 'Here, Charlie, here's your printouts. Where do you want them?' And he took one look and just about passed out. He called me and said, 'You know, Fred, maybe you should just print out hourly samples of the engineering data.'" Even for Charlie Hall it wasn't necessary to personally review *every* scrap of routine information.

"He had little rules to help ensure we'd get past as many problems as possible," Jack Dyer said. Hand in hand with the goal of keeping the spacecraft simple was keeping the mission objectives clear and focused. *Pioneers 6* through *9* were going to study the Sun and interplanetary medium; *Pioneers 10* and *11* were going to Jupiter, and *Pioneer Venus* was going to Venus. Period. Any alterations or additions to those simple objectives, Hall knew, would mean more work, greater complexity, tighter schedules, and most importantly, added expense. A PI might come up with a lovely idea for another experiment or task for Pioneer, but if it didn't harmonize perfectly with the basic mission objectives, no amount of pleading or arguing would sway Hall. "He would have considered it a great personal defeat to have had to expand greatly the projected cost because people in the course of development thought of better ways to do things," said Dyer.

Even when a proposal involved something that had absolutely no effect on the project budget or spacecraft design, Hall was tough to convince. When Dyer found a way to tweak *Pioneer 10*'s trajectory to allow it to occult (pass behind) Jupiter's moon Io, permitting some important observations to be made as an unexpected bonus of the Jupiter encounter, Hall resisted at first, worried about possible operational complications. "We had a hard time convincing Charlie that it was worth the small maneuver to pass behind Io," Dyer recalled with a chuckle. Later, Hall was also initially reluctant to pursue the opportunity to send *Pioneer 11* on to Saturn. Only when he was satisfied that the primary mission objectives were in no danger, or had already been met, would he relent.

Hall also granted an unusual degree of freedom to his prime contractor, with the understanding that existing technology was to be used whenever possible, either as is or as the foundation of modified design. Although the *Pioneer F/G* design was outwardly quite different from the earlier Pioneer spacecraft, it was a natural extension of those craft rather than a wholly new and untried concept. The existing technology rule extended to the PIs as well, who were strongly discouraged from designing instruments too radical or experimental in concept. "Hall was quite instrumental in preventing unproven components from getting onboard," John Simpson remembered with amusement.

Either personally or through his staff, Hall kept close tabs on all aspects of the project, whether at Ames, TRW, or other NASA centers and subcontractors. John Foster, Hall's immediate superior at Ames, said: "He knew every system on that thing [the Pioneer spacecraft]; his deputies were just extensions of himself."

Charlie Hall could be demanding, he could be difficult, he could be a stern taskmaster. Yet he was never a petty martinet. All who worked with him and for him regard him with the ultimate respect, admiration, even affection. "I think all of us were in total awe of him," Tom Gehrels remembered. "When Charlie Hall spoke, you would shut up. He was totally respected and maybe it's an improper word in the harsh environment of engineering, but a beloved person." Robert Kraemer wrote: "He obtained the facts, made timely decisions and assignments, and without raising his voice or pounding the table made it quietly clear when he expected actions to be completed."

As a leader who innately understood and appreciated the value of teamwork, Hall always firmly denied any claims that he was the chief architect of Pioneer's success. But the people he led beg to differ. It was his consummate skill in managing resources and people that allowed all the Pioneer spacecraft to not only fulfill their mission goals but surpass them in ways that made history.

DREAMS VERSUS REALITY

The enormous success of Pioneer and other space probes such as *Voyager* and *Galileo* raises an inevitable question, one that has been argued from the dawn of the space age. If automated spacecraft can do so much, why send human beings into space?

The original motivation was more political than scientific. If the Soviets were going to put men into space, we had to do likewise to prove to the world that we could compete with and surpass the achievements of the Communists. In 1961, when John F. Kennedy challenged NASA to put a man on the Moon and return him safely to Earth by the end of the decade, he fired the starting gun of a Cold War race that was more about asserting our national will and technological prowess than exploring the universe. Landing on the Moon was about national prestige and honor. It was about beating the Russians. In the final analysis, scientific discovery and the spirit of exploration were incidental dividends, not the motivating force that drove a nation to spend $40 billion in less than 10 years to accomplish the seemingly impossible. Although the Apollo program will abide as a supreme achievement of humankind, the rapid dwindling of its public and political support after it had achieved its basic goals demonstrates that it wasn't truly about science. As far as the serious scientific exploration of the Moon is concerned, the plug was rudely yanked just as things started to become interesting on the later Apollo missions, and human explorers haven't been back to the Moon since.

But the centuries-old dream of human spaceflight has a powerful allure. Though it's taken many different forms over the years, the dream includes some basic touchstones: routine flights to giant Earth-orbiting space stations; human colonies on the Moon and Mars; grand voyages of exploration throughout our solar system and later to the stars. At first the dream was held only by a handful of visionaries who were mostly ridiculed as naive crackpots (except when their governments decided to use them to build weapons). But slowly and surely, their vision spread throughout our culture to the point where it's now shared by millions. It's now so pervasive that it almost feels inevitable, like cosmic manifest destiny.

The dream doesn't go away easily, even when we've spent the decades since the last lunar landing venturing no farther into the universe than low Earth orbit. But it definitely lost a great deal of its luster with the hard experience of real human spaceflight. When the Space Shuttle was sold to us back in the 1970s, the blithe promise was that it would make spaceflight cheap and routine. It would fly almost every week and lift all sorts of payloads at a cost of only a few hundred dollars a pound, maybe even much less. After a while ordinary folks might be able to buy a ticket on the Space Shuttle for not much more than the cost of a trip to Paris. Needless to say, it didn't work out that way. The shuttle's been flying for only about 20 years and has barely made just over a hundred trips to orbit. A mission takes a year or more to plan and can cost over $1 billion. Rather than just a week or two, it takes many months to turn around a shuttle for its next flight. Because of the Space Shuttle's incredible complexity, thousands of people are needed to prepare for and launch each mission.

It's true that much of this is due to design compromises and budget cuts imposed on the program from its very beginnings. But the elemental reason for these problems is much simpler: flying into space is expensive. It's not a matter of inept administration or porkbarrel politics. It's simple physics. A vast amount of energy is needed to lift anything out of the Earth's gravity well and always will be. Generating that much energy costs money, a lot of it, and always will. There are no magical shortcuts, no ways to finesse the problem through a clever technical fix as Scotty always did on *Star Trek*.

Getting off the Earth is only the first hurdle. Space is an inherently hostile environment, and spacecraft must be constructed to withstand fierce radiation, enormous acceleration, extreme changes

in temperature and pressure, the lack of atmosphere, and microme-
teorite impacts. All these difficulties are greatly magnified when human beings enter space. We have to be given air to breathe, kept not
too warm and not too cold, shielded from radiation, and protected
from extreme acceleration that would crush our bodies. The technology needed to do all this adds great amounts of weight to a spacecraft—and every extra ounce of weight adds expense.

Added to all of these difficulties, of course, is the danger. No
matter what precautions and safety measures are taken, sending human beings into space is an inherently risky proposition. We've
grieved the loss of fourteen brave men and women with the *Challenger* and *Columbia* shuttle disasters. Much earlier, Gus Grissom, Ed
White, and Roger Chaffee perished in the *Apollo 1* fire, and Jim Lovell,
Fred Haise, and Jack Swigert came very close to losing their lives on
Apollo 13. Since the 1960s, the Russians have lost even more lives in
their own space program. The men and women who died for the
exploration of space knew the risks they were taking and willingly
accepted them, but the public and the politicians who send them
away from Earth seem less willing to do the same, demanding an
unattainable goal of absolute safety, questioning whether the loss of
human life is worth it.

If our objectives for going into space are purely practical or
scientific, the answer is probably *no*. Space travel has brought immea-
surable benefits to humanity, from providing instantaneous communications on a worldwide scale, to greatly increasing the accuracy of
weather forecasts, to giving us a new scientific and philosophical perspective on the universe. But those benefits, and indeed everything
that makes spaceflight essential to modern civilization, are achieved
by unmanned spacecraft. All of the practical value of our presence in
space is derived from satellites and space probes. James Van Allen, a
famous (or perhaps, in some circles, infamous) critic of human spaceflight, is adamant on this point. "The manned program has been no-
tably deficient in any important results of a scientific or practical
character. The utilitarian aspects of space technology such as tele-

communications, navigation, earth reconnaissance, study of the ocean, meteorology, all of these are done by unmanned spacecraft It's absurd to think of those being done by manned spacecraft."

Pioneer PI Robert Soberman takes a different view: "If you design a robot, it will only measure what you anticipate it will measure," he says. "If you get an answer that you don't expect, you're completely in the dark. Man has a much greater acceptance ratio for the unforeseen. As much as it costs, there's probably justification to have people out there."

Van Allen points out that NASA's emphasis on human spaceflight has usually been to the detriment of science. The Hubble Space Telescope, for example, was originally intended to be placed in a much higher orbit, but as Van Allen observed, "the astronomers were sort of co-opted by the manned spaceflight people to put it in an orbit where it could be man-tended. The Earth's in the way about half the time... and because of air drag they have to boost it up occasionally to keep it in orbit. The whole thermal problem is very difficult, going into shadow and back into sunlight, everything creaks and shudders and groans."

Other Pioneer veterans are also dubious about the utility of humans in space. "I've always questioned the scientific merit and return of manned spaceflight," Richard Fimmel said. "It's too expensive to keep us creatures alive. That's where all the money goes. If I could say what could have been done better in the past 20 or 30 years, it would be to put less emphasis on manned space." Political realities, however, dictated otherwise, as Fimmel remembered. "I talked to a congressman once and I told him how I felt about the manned space program. And he said, 'Mr. Fimmel, let's face it. If we didn't have Apollo, we'd have no NASA, because that's what the public understands. Putting a man in space is important to them. It's exciting to them. So rather than pick on it, be glad you've got it. I can get my constituents to vote money for NASA because of the manned space program, not because of the science program they don't understand.'"

If our machines are already reaping the practical benefits of space, is there any other reason to go there? "I think it's in the nature of man to want to investigate and learn more," says Fimmel. "And if we ever lose that urge, that desire to learn more, to expand our knowledge, we'll be a civilization that starts to decline. I think it's important. Not because of what we might benefit from it, but from all the things that it inspires."

So the question becomes that of who or what will and should do the exploring, human beings or automated spacecraft? If our purpose is to learn about and see the universe, does it matter, in the final analysis, whether we do so through our own eyes or those of our mechanical representatives?

Some scientists and engineers promise that when virtual reality technology is truly perfected and applied in unmanned spacecraft, the all-but-literal experience of standing on Mars or any other planet will be available to everyone, not just a small and highly select group of astronauts. True, it wouldn't be the same as actually being there, but will future generations, acclimatized to a world of pervasive and ever more realistic and interactive media, think so? To today's children who already spend so much of their time immersed in the virtual universes of the Internet and video games, the difference may be little more than superficial. Why spend the enormous sums necessary to physically put humans in space, subjecting them to all its dangers and discomforts, if they can see, feel, and do it all safely while still back on Earth?

In fact, it's already happened. In July 1997 *Mars Pathfinder* bounced to a landing on Mars and sent forth the wheeled Sojourner robot to explore the Martian landscape. The little robot captured the world's imagination as millions of people clogged the NASA Web site to watch its explorations for themselves in real-time images direct from the Martian surface. As journalist Marina Benjamin writes in her book *Rocket Dreams*, "*Pathfinder* gave Earth a first taste of what virtual space exploration might be like. Sojourner was our prosthetic eye: we saw what it saw as it panned across the rouged rubble of the Martian terrain." It also brought home the fact that, for the vast majority of human beings who would never have the chance to walk on another planet, there was no essential difference between seeing an other world through the lens of a camera operated by a robot and one operated by an astronaut who was actually there. And Sojourner's success was by no means a fluke. The phenomenon was repeated early in 2004 with the more sophisticated Spirit and Opportunity Mars rovers.

Virtual space exploration may be only the beginning. Suppose that a workable, direct interface between the human brain and the computer is someday realized, or even the literal transference of an individual human mind into a machine. Then we could go anywhere our machines can go, even into environments no human organism could venture. If our minds are there, physically embodied in a machine, does it matter that our frail human bodies have been left behind?

Maybe the final barrier to space exploration and colonization is that it's just too difficult for complex biological organisms to survive outside their original planetary environments. Maybe this is a universal, unalterable physical obstacle that extraterrestrial civilizations also face, so that it's literally not possible for short-lived and fragile biological organisms to conquer space, at least not outside their immediate celestial neighborhood. Maybe the only sort of spacecraft that ply the interstellar byways are machines, which don't have to worry about the cold or radiation or the vast time it takes to traverse the incredible distances between the stars. Maybe the only civilizations that manage to become truly space-faring are those that are technologically able to move beyond their biologically dictated limitations and into space either through machine proxies or by evolving into some sort of organic-machine hybrid.

And, of course, maybe such ideas are nonsense. There may yet be a stunning technological breakthrough that suddenly makes spaceflight cheap and affordable, making practical and economic concerns irrelevant. Legendary science fiction author Arthur C. Clarke, the man who invented the concept of communications satellites way back in

1945, once said: "When a noted scientist says that something is possible, he's almost certainly right. And when he says something is *impossible*, he's almost certainly wrong." But no such fabulous technological miracle is on the horizon. The cruel realities of Newtonian physics don't care about our glorious dreams of humans traversing the spaceways to the planets and, someday, the stars. Unappealing and unromantic as it may be to our grounded and thuddingly literal minds, the prospect of space exploration by virtual reality or artificially intelligent machines is much more likely, at least for the moment.

If this is indeed the future of exploration, then the Pioneer spacecraft have provided us with a glimpse of that future by proving that machines can go places where human creatures could never survive and show us sights that our own eyes could never see. They have demonstrated the art and science of remote exploration, how spacecraft can be made so sturdy and reliable that they can be made to exceed their limits by the ingenuity of humans billions of miles away. They have shown that we can literally extend our senses and our minds far beyond the boundaries of Earth and our own biological frailties.

We've always looked to the horizon, to the unknown lands beyond, feeling ourselves drawn there and knowing that eventually we would explore and tame those lands. Now, for the first time in human history, we have seen a horizon we can approach, an unknown eternity that we can observe from afar but possibly never reach and never touch. It's a bitter disappointment and a terrible affront to our ego, our belief that we can do anything and conquer any frontier. It will take some time for us to accept it, for the grieving process for what we dreamed could be ours to come to an end. But it may not be us in our flimsy biological forms that explore space; it may be our proxies, either machines or perhaps some even more exotic creation we have yet to imagine and invent, but that will still be artificial.

Pioneer 10 was the first. We can only hope and strive for it not to be the last. Because one day, about 5 billion years from now, our Sun will expand into a red giant star, consuming the Earth—and humanity itself, if we haven't moved elsewhere by then. When our sun and consequently our planet die, the Pioneers, Voyagers, and whatever other spacecraft may follow them into the galaxy may be all that are left to tell the universe that we were once here.

THE VOID

Charlie Hall never saw *Pioneer 10* reach its 30th birthday. He died of cancer on August 26, 1999, in a hospital in Mountain View, California, not far from his home and the Ames Research Center where he had spent his entire career. He was 79 years old. He left behind a legion of friends and colleagues who remember him warmly for his "intelligence, persistence and leadership throughout his career," as NASA administrator Dan Goldin phrased it in a memorial service. And, of course, he left behind the remarkable Pioneer spacecraft. Perhaps no other human being has such a unique legacy—a legacy that might be almost forgotten by now if not for the dogged efforts of the people Hall inspired and led.

Almost 2 $^1/_2$ years later, Hall's *Pioneer 10* followed him into silent oblivion. On January 22, 2003, the spacecraft was contacted by the DSN and responded, after a delay of 22 hours and 35 minutes. The signal was barely detectable, at the absolute threshold of the DSN's ability to pick it out of the cosmic static and too weak and faint to carry any usable telemetry from Van Allen's instrument. Another attempt at contact on February 7 was answered only by silence. Finally, after 31 years, *Pioneer 10*'s link with home had stretched too far and faded into nothingness. Engineers determined that the spacecraft's power had dropped to the point where the transmitter couldn't generate a signal strong enough to be heard across the more than 8 billion miles to Earth. No further efforts to try making contact were made.

Pioneer 10 had joined its sister, *Pioneer 11*, in eternal solitude. It was on its own, forever. It happened quietly, without any great fanfare or elaborate commemorative events such as those that had accompanied its departure from the solar system in 1983 and its 25th anniversary in 1997. The project veterans and scientists were saddened but not surprised. For them it was the close of a major chapter of their lives, one that for some had been the center of their careers. Although they had all moved on to other projects since the official end of the Pioneer mission, the awareness of *Pioneer 10*'s continued link with home had always been there in the background, a faint heartbeat of past glory. Now it was gone. Upon hearing the news, the controllers, engineers, scientists, and technicians shook their heads and marveled. Then they went back to work on their current jobs, new missions, new spacecraft, new discoveries, many based on the foundations built by and following the paths blazed by *Pioneer 10*.

Its work finished at last, its voice stilled, *Pioneer 10* drifts ever deeper into the galaxy. In about 26,135 years, it will pass slightly over 6 light years to the nearest star to the solar system, Proxima Centauri. Six thousand five hundred years later, it will pass within 4 light years of the star Ross 248. Perhaps someday it will become captured in the gravitational embrace of another star and find a new haven, orbiting in a solar system far from its home. Or it might encounter a stray asteroid, interstellar dust cloud, or other body that will destroy it. Most likely, *Pioneer 10* will just continue to drift farther into the universe, its antenna forever pointed back over the immense distance across which it had come, long after the star from whose influence it first departed, the planet that built and launched it, the worlds it first explored, have burned out and evaporated into cosmic debris. And then *Pioneer 10* will be carrying out its true and final mission: eternally marking the name of humanity on the depths of space.

Notes

Except where noted below, all comments in the text by individuals are from interviews with the author.

INTRODUCTION—
MESSAGE IN A BOTTLE

p. 5 "... by the time the probe . . .": United Press International release quoted in *Philadelphia Inquirer*, February 27, 1972.

p. 6 "Despite this planet's troubles . . .": "The Lonely Journey," *New York Times*, December 5, 1973.

1
EMBARKATION

p. 7 Ames early history: Glenn E. Bugos, *Atmosphere of Freedom: Sixty Years at the NASA Ames Research Center*; Elizabeth A.

Muenger, *Searching the Horizon: A History of Ames Research Center,*
1940-1976.

p. 15 "At Headquarters we were interested . . .": Oran W. Nicks,
Far Travelers: The Exploring Machines.

2
REACHING INTO THE VOID

p. 19 Details on the early Pioneer missions: William E. Bur-
rows, *Exploring Space: Voyages in the Solar System and Beyond* and
This New Ocean: The Story of the First Space Age; Homer E. Newell,
Beyond the Atmosphere: Early Years of Space Science; Asif A. Siddiqi,
*Deep Space Chronicle: A Chronology of Deep Space and Planetary
Probes 1958-2000.*

p. 22 *Pioneer V:* "U.S. Rocket Put into Sun Orbit"; "Pioneer V is
Launched"; "Pioneer V in Orbit"; *New York Times,* March 12-13, 1960.

p. 36 "I am shocked and disappointed . . .": Telegram from John
A. Simpson to Charles Hall, October 17, 1966; Letter from Simpson
to Dr. A. Schardt, October 18, 1966; John A. Simpson Papers, University of Chicago Regenstein Library Special Collections

p. 37 "*Pioneer 7* status report . . .": Telegram from Norman F.
Ness to Hall and NASA Headquarters, November 16, 1966, JAS Papers.

p. 37 "The Mariner encounter will affect . . .": Memo from G.
Lentz to Simpson et al., June 25, 1969, JAS Papers.

p. 37 "It is regrettable that . . .": Alois Schardt to John Simpson,
September 7, 1967, JAS Papers.

p. 39 *Pioneer 6* contact: NASA Ames Research Center press release, December 9, 2000.

3
SOMETHING MAN HAS NEVER DONE BEFORE

p. 40   Early mission failures: Burrows, *Exploring Space*; Siddiqi, *Deep Space Chronicle*.

p. 45   Flandro: "Fast Reconnaissance Missions to the Outer Solar System Utilizing Energy Derived from the Gravitational Field of Jupiter," April 18, 1966. Quoted in *Exploring the Unknown: Selected Documents in the History of the U.S. Civil Space Program*. Volume V: Exploring the Cosmos.

p. 46   Lassen and Park briefing: "Deep Space Probes: Sensors and Systems," Symposium on Unmanned Exploration of the Solar System, American Astronomical Society, February 8-10, 1965, Denver, Colorado. JAS Papers.

p. 47   Galactic Jupiter Probe: *Galactic Jupiter Probe*, NASA Goddard Space Center, February 1967. JAS Papers.

p. 49   "I made such a nuisance of myself . . .": James A. Van Allen, *Twenty-five milliamperes: A tale of two spacecraft,*" Journal of Geophysical Research vol. 101, no. A5, May 1, 1996.

p. 49   "Two exploratory probes . . .": Quoted in Burrows, *Exploring Space*, pp. 265-66.

p. 50   "Forget the Jupiter mission . . .": J.V. Foster, "History of Project Approvals for the Pioneer Programs," NASA Ames Research Center, March 1974.

4
THE SOLE SELECTION

p. 55   "Don, a true master . . .": Robert S. Kraemer, *Beyond the Moon: A Golden Age of Planetary Exploration 1971-1978*, p. 64.

p. 58   ". . . it was in the AEC's best interest . . .": Ibid., p. 67.

p. 62   NASA Announcement of Opportunity, June 10, 1968: JAS Papers.

p. 63   "Very little has been firmly decided . . .": Joe O'Gallagher to John A. Simpson, Memo, August 28, 1968. JAS Papers.

p. 64   "Unlike some of their colleagues . . .": John E. Naugle, *First Among Equals: The Selection of NASA Space Science Experiments*, NASA SP-4215.

p. 66   "I awaited NASA's formal decisions . . .": James A. Van Allen, *Twenty-five milliamperes*.

p. 69   "Tension was building . . .": Tom Gehrels, *On the Glassy Sea: An Astronomer's Journey*, pp. 125-26.

p. 70   "Deficiencies of design . . .": James A. Van Allen, *Twenty-five milliamperes*.

5
COUNTDOWN AND CONTROVERSY

p. 72   RTG moisture problem: Muenger, *Searching the Horizon*.

p. 73   ". . . not immune to puncture . . .": John Simpson to Joseph Lepetich, September 28, 1971. JAS Papers.

p. 79   "We're fascinated by . . .": *Birmingham News*, February 28, 1972. Quoted in *Pioneer to Jupiter: A History*, TRW Systems Group, November, 1973.

p. 79   "Despite the uncanny mastery . . .": "Message Found in Space," *New York Times*, March 4, 1972.

p. 79   "We do not know . . .": Frank Drake, *Time*, March 6, 1972.

p. 80   "When the plaque design . . .": Kraemer, *Beyond the Moon*, p. 75.

p. 80   "Uphold community standards": Quoted in William Poundstone, *Carl Sagan: A Life in the Cosmos*.

p. 81   Plaque reactions: "'Hey Look—It's a Bird!'": *Wall Street Journal*, May 23, 1972

p. 81   "I was shocked . . .": Quoted in Poundstone.

p. 81   "I am sure that . . .": Quoted in *Pioneer to Jupiter: A History*, TRW Systems Group, November, 1973.

p. 81   Cartoons: Ibid.

p. 82   "We didn't realize it . . .": Quoted in *Wall Street Journal*, May 23, 1972.

p. 82   "The message is for us too . . .": Quoted in Cynthia Ozick, "If You Can Read This, You Are Too Far Out," *Esquire*, January 1973.

p. 83   "We had already started . . .": Frank Drake, *Is Anyone Out There?: The Scientific Search for Extraterrestrial Intelligence.*

6
SPRING AT THE CAPE

p. 85   ". . . very disappointed . . .": Quoted in "Flight to Jupiter Is Set for Sunday," *New York Times*, February 25, 1972.

p. 85   ". . .becomes even more crucial. . .": Ibid.

p. 91   "Happy as a tick . . .": Quoted in "Pioneer Craft on Course," *New York Times*, March 4, 1972.

p. 93   Cable snag: Burrows, *Exploring Space.*

p. 94   "I caught a flight . . .": Gehrels, *On the Glassy Sea*, pp. 126-27.

p. 96   "Based on a variety of analyses . . .": NASA press release, July 14, 1972.

p. 98   "There's nothing surprising . . .": Quoted in "Pioneer Enters Asteroid Belt," *Ames Astrogram*, July 20, 1972.

7
TWELVE GENERATIONS FROM GALILEO

p. 101   "Had we used . . .": NASA press conference transcript, February 15, 1973.

p. 102   "If we had provided . . .": Quoted in "Jupiter's Pull Speeds Up Pioneer 10," *Aviation Week & Space Technology*, November 19, 1973.

p. 106   "I think most of the center personnel . . .": Kraemer, *Beyond the Moon*, p. 72.

p. 108   "This is an unusual event": Quoted in *Pioneer: First to Jupiter, Saturn, and Beyond,* Richard O. Fimmel, James Van Allen, Eric Burgess. NASA SP-446.

p. 109   "We are only twelve generations . . .": Ibid.

p. 110   Scientists camped out at Hall's door: Kraemer, *Beyond the Moon,* p. 73.

p. 112   "Misery of miseries . . .": Gehrels, *On the Glassy Sea,* pp. 128.

p. 112   ". . . a funeral parlor . . .": Quoted in Eric Burgess, *By Jupiter: Odysseys to a Giant,* p. 40.

p. 112   "The TRW engineers said flatly . . .": Kraemer, *Beyond the Moon,* p. 74.

p. 117   ". . . about the same amount of time . . .": Quoted in "Pioneer 10's Manager," *New York Times,* December 6, 1973.

p. 117   "We can say that we sent . . .": Quoted in *Pioneer: First to Jupiter, Saturn, and Beyond.*

p. 118   Nixon: Quoted in *New York Times,* December 9, 1973.

p. 118   Commander Dunning letter: "Message From the Men of Jupiter," *Ames Astrogram,* March 15, 1974.

8
FILLING IN THE GAPS

p. 120   "No further problems are expected": NASA press release, May 17, 1973.

p. 124   "The primary concern at NASA . . .": *New York Times,* December 9, 1973.

p. 128   Pioneer Direct Mode: "Pioneer Milestone," *Ames Astrogram,* June 6, 1974.

p. 131   DSN Canberra strike: *Pioneer: First to Jupiter, Saturn, and Beyond.*

p. 133   "Flew into the fiery jaws . . .": Quoted in "Pioneer Leaves Jupiter's 'Dragon,'" *New York Times,* December 4, 1974.

p. 135   Simpson article: "Journey to Jupiter," JAS Papers.

9
A JEWEL IN THE NIGHT

p. 140   "Precursor missions . . .": NASA press conference transcript, February 15, 1973.

p. 142   Broadfoot, Niehoff: Quoted in Frank Don Palluconi and Gordon H. Pettengill, editors, *The Rings of Saturn: Proceedings of the Saturn's Rings Workshop Held at JPL, July 31 and August 1, 1973*, NASA SP-343 (Washington, D.C., 1973), pp. 205-6.

p. 143   ". . .pass either through the dark space . . .": NASA press release, May 26, 1976.

p. 144   ". . .if we knew this crossing point . . .": John Casani to Thomas Young, February 3, 1977, JAS papers.

p. 144   "If Pioneer made it through . . .": Mark Washburn, *Distant Encounters: The Exploration of Jupiter and Saturn*, p. 164.

p. 144   "The opportunity to explore . . .": David Morrison to Charlie Hall, July 22, 1977, JAS Papers.

p. 145   "Certainly, *if* a trade must be made . . .": Ibid.

p. 145   "We do not find agreement": Charlie Hall to Pioneer Principal Investigators, September 20, 1977 JAS Papers

p. 146   "A *Science News* poll . . .": Jonathan Eberhart, "The Great Saturn Quandary," *Science News*, October 15, 1977.

p. 147   "The Voyager Project's preference . . .": Charlie Hall to Thomas Young, November 8, 1977, JAS Papers.

p. 147   "The predictable science return . . .": *Report of Pioneer 11 Targeting Meeting at Ames Research Center, November 1, 1977*, JAS Papers.

p. 148   Young letter: Thomas Young to Charles Hall, December 1, 1977, JAS Papers.

p. 150   "The intensity of conjecture grew . . .": Van Allen, *Twenty-five milliamperes.*

p. 152   "There were scattered cheers . . .": David Morrison, *Voyages to Saturn*, NASA SP-451, p. 24.

p. 153   "At the next morning's group meeting . . .". Van Allen, *Twenty-five milliamperes.*

p. 153   "Well, if it's like everything else . . .": NASA press conference transcript, December 4, 1974.

p. 153   Wolfe/Van Allen bet: Van Allen, *Twenty-five milliamperes.*

10
PLANET OF CLOUDS

p. 163   "Hughes would be content to break even . . .": Kraemer, *Beyond the Moon*, pp. 211-12.

p. 164   "The only suitable material . . .": Ibid., p. 214.

p. 166   "Fortunately, Charlie Hall's hair was . . .": Ibid., p. 215.

p. 173   "A rain of science . . .": "Venus Probes Yield Mass of Data," *New York Times*, December 10, 1978.

11
WHISPERS ACROSS THE ABYSS

p. 183   "*Pioneer 10* has moved past the orbit . . .": John A. Simpson, James Van Allen, Frank B. McDonald to Ichtiaque Rasool, May 18, 1976, JAS Papers.

p. 186   "For most of their long lives . . .": Douglas J. Mudgway, *Uplink-Downlink: A History of the Deep Space Network*, NASA SP-2001-4227, p. 331.

p. 188   "We strongly urge NASA . . .": "Draft Resolution for the NASA Interplanetary Sciences Working Group," February 7, 1977, JAS Papers.

p. 188   "Why Is It Vital for NASA to . . .": Simpson memo to Richard Fimmel, December 4, 1981.

p. 188   "One of the most incredible features . . .": James Van Allen, "Pioneer's Unfunded Reach for the Stars," *Aviation Week & Space Technology*, April 12, 1982.

p. 189   "There were celebrations . . .": "As Pioneer Flies On," *Aviation Week & Space Technology*, June 20, 1983.

p. 190   "Tomorrow, on to the stars": Quoted in "Spacecraft Leaves Realm of Planets," *New York Times*, June 14, 1983.

p. 192   Pioneer Uranus, Out-of-Ecliptic, Jupiter Orbiter Probe studies: JAS Papers.

p. 200   "*Pioneer 10* exemplifies . . .": Quoted in NASA press release, February 27, 1997.

12
LONE SURVIVOR

p. 209   "As an eternal optimist . . .": Quoted in NASA press release, March 4, 2002.

p. 209   "The electronics on *Pioneer 10* . . .": Quoted in "Pioneer 10 Leaves Known Solar System," *Aviation Week & Space Technology*, June 20, 1983.

p. 213   "At the peak of the project . . .": *New York Times*, December 6, 1973.

p. 216   "He knew every system . . .": Quoted in Muenger, *Searching the Horizon*, p. 219.

p. 216   "He obtained the facts . . .": Kraemer, *Beyond the Moon*, p. 65.

p. 221   "Pathfinder gave Earth a first taste . . .": Marina Benjamin, *Rocket Dreams*, p. 176.

p. 224   Hall death, Goldin quote: "Famed NASA Pioneer Project Manager Charles Hall Dead at 79," *Ames Astrogram*, September 13, 1999.

Bibliography

Benjamin, Marina. *Rocket Dreams.* New York: Free Press, 2003.

Bugos, Glenn E. *Atmosphere of Freedom: Sixty Years at the NASA Ames Research Center.* NASA SP-4314. Washington, D.C.: Government Printing Office, 2000.

Burgess, Eric. *By Jupiter: Odysseys to a Giant.* New York: Columbia University Press, 1982.

Burrows, William E. *Exploring Space: Voyages in the Solar System and Beyond.* New York: Random House, 1990.

————. *This New Ocean: The Story of the First Space Age.* New York: Random House, 1998.

Drake, Frank. *Is Anyone Out There?: The Scientific Search for Extraterrestrial Intelligence.* New York: Delacorte Press, 1992.

Fimmel, Richard O., James Van Allen, Eric Burgess. *Pioneer: First to Jupiter, Saturn, and Beyond.* NASA SP-446. Washington, D.C.: Government Printing Office, 1980.

————, Lawrence Colin, Eric Burgess. *Pioneering Venus: A Planet Unveiled.* NASA SP-518. Washington, D.C.: Government Printing Office, 1995.

Gehrels, Tom. *On the Glassy Sea: An Astronomer's Journey.* New York: American Institute of Physics/Springer-Verlag, 1988.

Heppenheimer, T.A. *Countdown: A History of Space Flight.* New York: John Wiley & Sons, 1997.

Klaes, Larry. "The Robot Explorers of Venus, Part II." *Quest* 8-2 (2002), pp. 24-30.

Kraemer, Robert S. *Beyond the Moon: A Golden Age of Planetary Exploration 1971-1978.* Washington, D.C.: Smithsonian Institution Press, 2000.

Logsdon, John M., ed. *Exploring the Unknown: Selected Documents in the History of the U.S. Civil Space Program.* Volume V: Exploring the Cosmos. NASA SP-2001-4407. Washington, D.C.: Government Printing Office, 2001.

Morrison, David. David Morrison, *Voyages to Saturn.* NASA SP-451. Washington, D.C.: Government Printing Office, 1982.

Mudgway, Douglas J. *Uplink-Downlink: A History of the Deep Space Network.* NASA SP-2001-4227. Washington, D.C.: Government Printing Office, 2001.

Muenger, Elizabeth A. *Searching the Horizon: A History of Ames Research Center, 1940-1976.* NASA SP-4304. Washington, D.C.: Government Printing Office, 1985.

Naugle, John E. *First Among Equals: The Selection of NASA Space Science Experiments.* NASA SP-4215. Washington, D.C.: Government Printing Office, 1991.

Newell, Homer E. *Beyond the Atmosphere: Early Years of Space Science.* NASA SP-4211. Washington, D.C.: Government Printing Office, 1980.

Nicks, Oran. *Far Travelers: The Exploring Machines.* NASA SP-480. Washington, D.C.: Government Printing Office, 1985.

Poundstone, William. *Carl Sagan: A Life in the Cosmos.* New York: Henry Holt and Company, 1999.

Reichhardt, Tony. "Gravity's Overdrive." *Air & Space Smithsonian,* February/March 1994, pp. 72-78.

Siddiqi, Asif A. *Deep Space Chronicle: A Chronology of Deep Space and Planetary Probes 1958-2000.* NASA SP-2002-4524. Washington, D.C.: Government Printing Office, 2002.

Soberman, Robert K., Maurice Dubin. *Dark Matter Illuminated.* Haverford, Pa.: Infinity Publishing, 2001.

Verschuur, Gerrit L. "Race to the Sun's Edge." *Air & Space Smithsonian,* April/May 1993, pp. 24-30.

Washburn, Mark. *Distant Encounters: The Exploration of Jupiter and Saturn.* New York: Harcourt Brace Jovanovich, 1983.

Wolverton, Mark. "Pathfinding the Rings: The Pioneer Saturn Trajectory Decision." *Quest* 7-4 (2000), pp. 5-11.

_____. "The Spacecraft That Will Not Die." *American Heritage of Invention & Technology,* Winter 2001, pp. 46-58.

_____. "30 and Counting." *StarDate.* March/April 2002, pp. 16-19.

INTERVIEWS

Burgess, Eric. December 4, 2003.

Campo, Ricardo. May 19, 2000.

Dyer, Jack. April 18, 2000.

Eggers, Alfred E. May 19, 2003.

Fimmel, Richard O. May 20, 2003.

Gehrels, Tom. July 8, 2003.

Hall, Charlie. August 13, 1999.

Jackson, Bob. August 13, 1999.

Lasher, Larry. August 12, 1999; April 7, 2000.

Lozier, Dave. May 17, 2000; December 6, 2001.

McDonald, Frank. July 9, 2003.

McKibben, Bruce. July 28, 1999.

O'Brien, Bernard. May 17, 2000.

Ryan, Bob. May 18, 2000.

Simpson, John A. July 15, 1999; August 5, 1999.

Soberman, Robert. January 15, 2003.

Tuzzolino, Anthony. July 15, 1999.

Van Allen, James. May 3, 2000; April 29, 2003.

Wirth, Fred. July 1, 2003.

Index

A

Advanced Research Projects Agency, 19–20
Allen, Harvey, 9, 12, 50
Alvarez, Luis, 111
American Astronomical Society, 46, 76
Ames Aeronautical Laboratory, 7–8
Ames Research Center, 4, 11, 18, 25, 33, 68, 73, 87, 93, 98, 105–108, 116, 122–123, 128, 131–133, 138, 143, 146, 149–152, 156, 160, 179, 193, 200, 203, 207, 216
  Public Affairs Office, 118
  Space Projects Facility, 38
  Vehicle Environment Division, 12–14
  See also Ames Aeronautical Laboratory
Anderson, John, 190
Announcements of Opportunity (AOs), 27, 62–63

Antenna repointing command sequence, 207
AOs. See Announcements of Opportunity
Apollo 1, 219
Apollo 11, 37
  command module, 191
Apollo 13, 219
Apollo program, 44, 90, 118
Aquila the Eagle, 199
Army Ballistic Missile Agency, 19
Asteroid belt, 42, 48, 100, 142
  entering, 95–99
Asteroid-meteoroid detector, 95
Atlas-Centaur launch vehicle, 59, 84–85, 87–90, 167
Atomic Energy Commission, 57, 71, 87, 213
Auer, Siegfried, 196
Aviation Week & Space Technology, 189

B

Bane, Don, 76
Beggs, James, 189
Bendix, 26
Benjamin, Marina, 221
Beyond the Moon, 54
Birmingham News, 79
Blunt-body concept, 12
Borman, Frank, 31
Broadfoot, Lyle, 142
Burgess, Eric, 75–76, 83, 108

C

California Institute of Technology, 15
Campo, Ric, 38–39, 150, 205, 207
Canopus, 91
Cape Canaveral, 11, 31, 83
Cape Kennedy, 31, 72, 75, 84, 98, 167
Casani, John, 144
Cassini spacecraft, 38
Chaffee, Roger, 219
Challenger disaster, 178, 195, 219
Charged particle experiment, 73, 92
Chicago Sun-Times, 80
Christian Science Monitor, 75
Clarke, Arthur C., 6, 222
Clementine spacecraft, 203
Columbia disaster, 219
Comet Halley, 176
Command sequences, antenna
    repointing, 207
Communications difficulties, 48, 102,
    184–191
Communications satellites, 222
Competition, 61–70
Compromise, 61
Computer use, 125–128, 205–206
    software upgrades, 206
Congress, 188
Contracts, cost-plus-incentives, 31

Controversy, 71–83
    message sent, 75–83
Cornell University, National
    Astronomy and Ionosphere
    Center, 76
Cortright, Edgar "Ed," 15, 25
Cosmic ray telescope, 181
Cost-plus-incentives contracts, 31

D

Deep Space Network (DSN), 33, 35, 37,
    39, 61, 93, 102, 120–123, 127,
    131–133, 152, 156, 170, 182,
    185–188, 199–208, 224
Defense Department, Advanced
    Research Projects Agency, 19–20
DeFrance, Smith "Smitty," 10–11, 16–18
Distant Encounters, 144
Dixon, William, 56, 61, 113
Doose, Lyn, 180
"Doose correction," 180
Doppler shift data, 91, 102, 190
Drake, Frank, 76–82
Dryden, Hugh, 15
DSL lines, 103
DSN. *See* Deep Space Network
Dubin, Maury, 196–197
Dunning, J.P., 118–119
Dyer, Jack, 73, 111, 146, 148, 190–192,
    211, 214–215

E

Edwards Air Force Base, 7–8
Eggers, Alfred "Al," 9, 12–17, 25–26, 54,
    156, 214
Einstein, Albert, General Theory of
    Relativity, 190
Esquire, 82
European Space Agency, 194

Explorer I, 25
Explorer VII, 22

F

Far Travelers, 16
Federation Aeronautique
    Internationale, 118
Fimmel, Richard, 106, 108, 180, 187–
    188, 199, 212–213, 220–221
Flandro, Gary, 45
Fletcher, James, 87–88, 137
Flux-gate magnetometers, 129
Foster, John, 161–162, 216

G

"Galactic Jupiter Probe," 47, 49
Galileo, Galilei, 109, 134
Galileo spacecraft, 4–5, 38, 178, 195, 217
Gegenschein, 92
Gehrels, Tom, 60, 67–69, 94, 109, 112,
    124, 211–212, 216
    imaging photo polarimeter, 60, 92–
    95, 103–104, 106, 150–151, 155,
    180, 196
Geiger tube telescope, 92, 181, 187,
    201–202, 208
Gemini 6, 31–32
Gemini 7, 31–32
General Theory of Relativity, 190
Goddard Space Flight Center, 10, 22,
    37, 47–50, 67
Goldin, Daniel, 4–5, 178, 203, 224
Goody, Richard, 160
"Grand Tour" idea, 45–46, 49, 85, 118,
    140, 192
Gravity assist technique, 45
"Great Galactic Ghoul," 41–42, 51, 94,
    157
Grissom, Gus, 219

H

Haise, Fred, 219
Hall, Charles Frederick "Charlie," 4–5,
    9, 12–19, 25–31, 34, 36, 46–47,
    50–58, 61–62, 67–70, 75, 84, 88,
    94–95, 98, 101, 107–117, 121–
    123, 131, 139, 144–148, 151–
    153, 156–157, 161–162, 166,
    180, 195, 200, 210–216, 224
    management style of, 68, 75, 213–
    217
Hayden Planetarium, 102
Hearth, Don, 54–55
Heliopause, 181, 190
Heliosphere, 188
Helium-vector magnetometers, 129
Hoagland, Richard, 76
Hollywood, 81
Holtzclaw, Ralph, 85, 87
Houston's Manned Spaceflight Center,
    106
Hubble Space Telescope, 220
Hughes Aircraft, 26–27, 162–163
Human spaceflight, dream of, 217–224
Hunten, Donald, 160
Huntress, Wesley, 200
Huntsville, 11

I

IGY. *See* International Geophysical
    Year
Imaging photo polarimeter (IPP), 60,
    92–95, 103–104, 115–116, 150–
    151, 155, 196
    Pioneer Image Converter System
    for, 106
*IMP (Interplanetary Monitoring
    Platform)* satellites, 67
Ingenuity, 223
Innovation, 40–51

International Astronomical Union, 175
International Geophysical Year (IGY),
    27
International Quiet Sun Year (IQSY),
    27
Interstellar meteoroids, 197
IPP. *See* Imaging photo polarimeter
IQSY. *See* International Quiet Sun Year

Khinoy, Andrew, 80
Kinard, William, 98, 102, 137
    meteoroid detection experiment,
    95, 196
Kitt Peak Observatory, 142
Kohlhase, Charles, 146
Kraemer, Robert, 54–55, 58, 80, 106,
    117, 133, 164, 216

J

Jet Propulsion Laboratory (JPL), 10,
    15, 18, 33, 45, 49, 106, 121–128,
    142–144, 149–150, 156, 185,
    190, 192, 199, 210, 213
Jodrell Bank radiotelescope, 24
JOP. *See* Jupiter Orbiter-Probe
Journal of Geophysical Research, 196
JPL. *See* Jet Propulsion Laboratory
Jupiter, 43–44, 49–50
    first probes of, 4
    gravitational field of, 100–101, 117,
    193
    magnetic field of, 104, 190
    moons of, 134, 136
    new knowledge about, 134–137
    particles and fields of, 44
    Pioneer 10 to, 100–119
Jupiter mission findings, 138–158
    future missions, 140–149
    going beyond Saturn, 156–158
    initiatives involving Saturn, 149–153
    new facts about Saturn, 154–156
Jupiter Orbiter-Probe (JOP), 194

K

Kennedy, John F., 44, 217
Kennedy Space Center, Vehicle
    Assembly Building, 165

L

Langley Aeronautical Laboratory,
    9–11
Lasher, Larry, 38–39, 202–203, 206–
    207, 209
Lassen, Herb, 46–47, 55–58, 61, 75, 91
Launch phase, 84–99
    entering the asteroid belt, 95–99
Launch window, 87
Lepetich, Joseph, 67, 73
Letters of Inquiry, 63
Lewis Research Center, 10
Lindbergh, Charles, 191
Los Angeles Herald-Examiner, 76
Los Angeles Times, 81
Lovell, James, 31, 219
Lovell, Sir Bernard, 24
Low, George, 162
Lowell, Percival, 1–2
Lozier, Dave, 38–39, 60–61, 86–87, 90,
    207–209
LPMB. *See* Lunar and Planetary
    Missions Board
Luna 1, 21
Luna 3, 21
Lunar and Planetary Missions Board
    (LPMB), 49
Lunar and Planetary Programs
    division, 16, 54
Lunar Prospector mission, 203

M

Magellan spacecraft, 175, 177–178
"Magnetic cleanliness," 28
Magnetodiscs, 113–114
Magnetometers, 59
    flux-gate, 129
    helium-vector, 129
Magnetosphere boundary, 105
Male chauvinism, 82
Manhattan Project, 28
Mariner 1, 40
Mariner 2, 40, 159
Mariner 4, 40
Mariner 9, 37
Mariner flights, 35
Mark, Hans, 108, 167
Mars
    contacting, 1–2
    Mariner flights to, 35
Mars Observer mission, 5
Mars Pathfinder mission, 5, 221
Mars Polar Lander mission, 5
Marshall Space Flight Center, 10
Martin Marietta, 162
McDonald, Frank, 105, 115, 145, 147,
    183
    cosmic ray telescope, 181
Messages
    carried by *Pioneer 10* to Jupiter, 3
    from the unknown, 1–6
Meteoroid detection experiment, 95
Milestones of Flight Hall, 191
Mini-"Grand Tour" idea, 192
Minovitch, Michael, 45
Mitchell, Jesse, 14–15
Moon shots, 20
Morrison, David, 144–145, 152
Mudgway, Douglas J., 186

N

NACA. *See* National Advisory
    Committee on Aeronautics
National Academy of Sciences, Space
    Science Board, 49
National Academy of Television Arts
    and Sciences, 107
National Advisory Committee on
    Aeronautics (NACA), 7, 9–10
National Aeronautics and Space
    Administration (NASA), 4, 9–
    18, 22, 24, 28, 30–31, 58, 76, 87,
    96, 105, 111, 117–118, 122, 137,
    140–148, 160, 163, 167, 177–
    184, 188, 192–200, 208, 216–
    217, 220
    Ames Research Center, 4, 11, 18, 25,
        33, 68, 73, 87, 93, 98, 105–108,
        116, 122–123, 128, 131–133,
        138, 143, 146, 149–152, 156, 160,
        179, 193, 200, 207, 216
    budget of, 18, 27, 44, 176
    Cape Canaveral, 11, 31, 83
    Cape Kennedy, 31, 72, 75, 84, 98
    Deep Space Network, 33, 35, 37, 39,
        61, 93, 102, 120–123, 127, 131–
        133, 152, 156, 170, 182, 185–188,
        199–208, 224
    Distinguished Service Medal, 180
    fiefdoms within, 156
    Goddard Space Flight Center, 10,
        22, 37, 47–50, 67
    higher authorities in, 13, 53
    Huntsville, 11
    Jet Propulsion Laboratory, 10, 15,
        18, 33, 45, 49
    Langley Aeronautical Laboratory,
        9–11
    Lewis Research Center, 10
    Lunar and Planetary Programs
        division, 16, 54

Marshall Space Flight Center, 10
Office of Space Sciences, 27, 62
Particles and Fields science
    committee, 14
policy of, 141
space sciences branch, 16–17
Space Technology Laboratories, 4,
    16, 26
See also National Advisory
    Committee on Aeronautics
National Aeronautics Association, 118
National Air and Space Museum, 194,
    200
Milestones of Flight Hall, 191
National Astronomy and Ionosphere
    Center, 76
National Oceanic and Atmospheric
    Administration, 34
Natural History, 81
Naugle, John, 64, 80, 140
Neff, Donald, 41
Ness, Norman, 37
New York Times, 6, 22–23, 79, 85, 117,
    124, 173, 213
Nicks, Oran, 15–16
Niehoff, John, 142
Nixon, Richard, 118
Nunamaker, Skip, 161

O

O'Brien, Bernard J., 55–56, 60–61, 90–
    91, 94, 97, 116–117, 209, 211
Office of Space Sciences, 27, 62
O'Gallagher, Joe, 63
*OGO-5. See Orbiting Geophysical
    Observatory 5*
Opportunity spacecraft, 222
*Orbiting Geophysical Observatory 5
    (OGO-5),* 73
Ozick, Cynthia, 82

P

PAET. *See* Planetary Atmosphere
    Experiments Test
Palomar Schmidt Telescope, 94
Park, Robert, 46–47
Particles and Fields science committee,
    14
Periapsis, 111
Philadelphia Inquirer, 80
Physical Review, 64
Pickering William, 15, 122
PICS. *See* Pioneer Image Converter
    System
Pioneer program, 18–19, 24–25, 53, 73,
    79, 96, 108, 123, 146, 157–158,
    169–170, 176, 179–201, 203,
    212, 217
alternatives, 195–198
communications difficulties, 184–
    191
costs of, 27
dealing with magnetism, 28
dealing with radiation, 29
design of, 209–212
extensions, 180–183
"luck" of, 153, 165
Mission Operations Control
    Center, 128, 152, 170, 190, 201,
    205, 207
missions of, 36
official closing, 198–201
options not implemented, 191–195
plaque for, 3, 76–83
reliability issues, 30
See also Principal investigators
Pioneer 0, 20
Pioneer 1, 20
Pioneer 2, 20
Pioneer 3, 20–21
Pioneer 4, 21
Pioneer 5/V, 22–25
Pioneer 6, 4, 31–33, 39, 62, 179, 210

Pioneer 6 through *9*, 29, 36, 40, 49, 53–54, 56, 126, 193, 210, 215
 endurance of, 35
Pioneer 7, 33, 87, 179
Pioneer 8, 33, 179
Pioneer 9, 33, 36–37, 97, 179
Pioneer 10 to Jupiter, 2–6, 34, 49, 52–59, 67–68, 72–75, 78–79, 82–84, 87–130, 136–141, 157, 162, 169, 173–174, 179–185, 188–194, 198–215, 223–225
 link to maintained, 202–225
 dream of human spaceflight, 217–224
 final silencing, 224–225
 Hall's management style, 213–217
 legacy of Pioneer's design, 209–212
 pressure to continue mission, 203–209
 message carried by, 3
Pioneer 11, 4, 49, 59, 67–68, 70, 74–75, 78, 98–99, 102, 117–121, 124, 128–143, 146–157, 162, 174, 179–194, 198, 215, 224
Pioneer 12, 167, 179
Pioneer 13, 167, 179
Pioneer Saturn, 137
Pioneer Venus, 161–164, 171–179, 194–195, 210, 215
Pioneer Image Converter System (PICS), 107, 116
PIs. *See* Principal investigators
Planetary Atmosphere Experiments Test (PAET), 160
Plaque, for Pioneer program, 3, 76–83
Plasma analyzer, 139, 150
Plutonium-238, 83
Principal investigators (PIs), 28, 33, 35, 64–71, 109, 111–112, 119, 121, 145–147, 187, 190, 195, 200, 211, 215, 220

Problem-solving, 120–137
 computer use, 125–128
 new knowledge about Jupiter, 134–137
 Pioneer 11 to Jupiter, 128–134
Proxima Centauri, reaching, 2, 225

R

Racism, 82
Radiation, dealing with, 29, 43, 74, 112, 115, 181
"Radiation death," 74
Radioisotope thermoelectric generators (RTGs), 57–58, 63, 66, 71–75, 83, 87–88, 98–99, 120, 130, 157, 198, 211
 plutonium dioxide capsules, 187
 transit satellite, 67
Rasool, Ichtiaque, 182
Reentry, surviving, 9
Relativity, General Theory of, 190
Reliability issues, 30
Requests for Proposals, 54
Reusable space shuttles, 46
Robinson Crusoe, 1
Rocket Dreams, 221
Ross 248, 225
RTGs. *See* Radioisotope thermoelectric generators

S

Sagan, Carl, 76–82
Sagan, Linda Salzman, 77, 80–82
Saturn
 first probes of, 4
 going beyond, 156–158
 initiatives involving, 149–153
 new facts about, 154–156
 rings of, 142–144, 155

Schardt, Alois, 37–38
Schirra, Wally, 31
Schlesinger, James, 88
Science, 196
Science News, 146
Seamans, Robert, 17–18
Search for Extraterrestrial Intelligence
    (SETI) Institute, 76, 204
Selection, 52–70
    competition in, 61–70
Selkirk, Alexander, 1
SETI. *See* Search for Extraterrestrial
    Intelligence
Simpson, John, 28, 36–38, 63, 69, 71,
    73, 96, 121, 132, 135, 145, 158,
    183, 187, 199
    charged particle instrument, 92
"Sisyphus" telescope, 95, 115, 139, 196
Smithsonian Institution, National Air
    and Space Museum, 190–191,
    194, 200
SNAP-19 (Systems for Nuclear
    Auxiliary Power), 57–58
Soberman, Robert, 98, 115, 195, 197–
    198, 220
    "Sisyphus" telescope, 95, 115, 139,
    196
Software upgrades, 206
Sojourner spacecraft, 222
Solar cells, 58, 63, 66
Solar flares, 35
Solar panels, 42–43
Solar probes, 16–17
Solar storms, 97
Solar wind, 97
Sonett, Chuck, 14, 16–17
Space exploration, 19–39
    checking in, 38–39
    risks inherent in, 219
    wide-open frontier of, 12
    workability issues, 24–38
Space Projects Facility, 38

Space Science Board (SSB), 49, 160
Space sciences branch, 16
Space Shuttle, 85, 177–178, 218
    reusable, 46
Space Technology Laboratories, 4, 16,
    22, 26
Space weather network, 34
Spaceflight, dream of human, 217–224
Speed of light, 43
Spin scan imaging, 60
Spin stabilization, 30
Spirit of Saint Louis, 191
Spirit spacecraft, 222
SSB. *See* Space Science Board
Stabilization
    spin, 30
    three-axis, 30, 149
"Stand-up" meetings, 110
Stanford University, 14
Star Trek, 218
Swigert, Jack, 219
Systems for Nuclear Auxiliary Power.
    See SNAP-19

T

Teledyne Corporation, 57, 71–72
Thiokol rocket stage, 90
Thor-Able rocket booster, 22
Three-axis stabilization, 30, 149
Time magazine, 41, 79
Trajectories, precise, 41
Traveling wave tube (TWT), 120
TRW, 17, 26–31, 46, 53–56, 61–62, 69–
    73, 84, 90–91, 112–116, 126,
    156, 162, 178, 193, 200, 209–213,
    216
    See also Space Technology
    Laboratories
TWT. *See* Traveling wave tube

U

United Press International, 5
University of Chicago, 63, 135
Unknown, messages from, 1–6
*Uplink-Downlink: A History of the Deep
  Space Network,* 186

V

Van Allen, James, 28, 36, 48–49, 63, 66–
  70, 105, 109, 113–114, 119–122,
  130–132, 145, 150, 153–154,
  158, 181–183, 187–189, 192–
  194, 199–204, 209–211, 220
Geiger tube telescope, 92, 181, 187,
  201–202, 208, 224
Van Allen radiation belts, 13, 20, 23
Vanguard mission, 19
VED. *See* Vehicle Environment
  Division
Vehicle Assembly Building, 165
Vehicle Environment Division (VED),
  12–14
Venera 4, 40
Venera 7, 160
Venus, 159–178
  atmospheric composition of, 174
  engineering factors, 161–168
  last Pioneer mission, 177–178

Mariner flights to, 35–36
new facts about Venus, 173–177
Pioneer 12 to Venus, 168–173
surface of, 172
Venus—Strategy for Exploration, 160
Viking missions, 182
Von Braun, Wernher, 19–21, 127
Voyager missions, 122, 125, 139–148,
  152, 188, 191–192, 210, 217

W

Wall Street Journal, 80–81
Washburn, Mark, 144
Westinghouse, 26
White, Ed, 219
White Sands, New Mexico, 165
Wilford, John Noble, 85, 173
Wind tunnels, 7–8
Wirth, Fred, 107, 109, 123, 125–126,
  132, 172, 202, 210–211, 214–215
Wolfe, John, 111, 150, 153
Wollam, Earl, 22
Workability issues, 24–38
Wright Brothers, 191

Y

Young, Thomas, 144, 146–149